T0214699

Synthesis Lectures on Learning, Networks, and Algorithms

Series Editor

Lei Ying, ECE, University of Michigan–Ann Arbor, Ann Arbor, USA

The series publishes short books on the design, analysis, and management of complex networked systems using tools from control, communications, learning, optimization, and stochastic analysis. Each Lecture is a self-contained presentation of one topic by a leading expert. The topics include learning, networks, and algorithms, and cover a broad spectrum of applications to networked systems including communication networks, data-center networks, social, and transportation networks.

Minghua Chen · Sid Chi-Kin Chau

Online Capacity Provisioning for Energy-Efficient Datacenters

 Springer

Minghua Chen
School of Data Science
City University of Hong Kong
Kowloon, Hong Kong

Sid Chi-Kin Chau
School of Computing
Australian National University
Canberra, ACT, Australia

ISSN 2690-4306   ISSN 2690-4314 (electronic)
Synthesis Lectures on Learning, Networks, and Algorithms
ISBN 978-3-031-11551-6   ISBN 978-3-031-11549-3 (eBook)
https://doi.org/10.1007/978-3-031-11549-3

This Springer imprint is published by the registered company Springer Nature Switzerland AG
The registered company address is: Gewerbestrasse 11, 6330 Cham, Switzerland

Acknowledgements

We would like to thank a number of persons who helped us to publish this monograph. First, we would like to thank publisher Michael Morgan for his kind assistance in the publication process. We are grateful to editors R. Srikant and Lei Ying for providing us this opportunity. We also would like to thank Xiaojun Lin and the anonymous reviewer for their very helpful comments.

May 2022
Minghua Chen
Sid Chi-Kin Chau

Contents

About the Authors

Minghua Chen received his B.Eng. and M.S. degrees from the Department of Electronic Engineering at Tsinghua University. He received his Ph.D. degree from the Department of Electrical Engineering and Computer Sciences at University of California, Berkeley. He was with Microsoft Research Redmond and The Chinese University of Hong Kong before joining School of Data Science, City University of Hong Kong, where he is a Professor. Minghua received the Eli Jury award from UC Berkeley in 2007 (presented to a graduate student or recent alumnus for outstanding achievement in the area of Systems, Communications, Control, or Signal Processing) and The Chinese University of Hong Kong Young Researcher Award in 2013. He also received several best paper awards, including IEEE ICME Best Paper Award in 2009, IEEE Transactions on Multimedia Prize Paper Award in 2009, and ACM Multimedia Best Paper Award in 2012. He served as Associate Editor of IEEE/ACM Transactions on Networking in 2014–2018. He received the ACM Recognition of Service Award in 2017 for the service contribution to the research community. He is currently a Senior Editor for IEEE Systems Journal (2021–present) and an Executive Committee member of ACM SIGEnergy (2018–present). Minghua's recent research interests include online optimization and algorithms, machine learning in power system operations, intelligent transportation systems, distributed optimization, delay-constrained network coding, and capitalizing the benefit of data-driven prediction in algorithm/system design. He is a Distinguished Member of ACM.

Sid Chi-Kin Chau received a Ph.D. degree from University of Cambridge with a scholarship by the Croucher Foundation Hong Kong, and a B.Eng. (First-class Honours) degree from the Chinese University of Hong Kong. He is with the School of Computing at the Australian National University. His research interests are related to computing algorithms and systems for smart sustainable cities, including smart grid, smart buildings, intelligent vehicles, and transportation. He also researches in broad areas of algorithms, optimization, Internet of Things, and blockchain. He was an Associate Professor with the Department of Computer Science at Masdar Institute, which was created in collaboration with MIT, and is a part of Khalifa University. Previously, he was a visiting professor at MIT, a senior research fellow at A*STAR in Singapore, a Croucher Foundation research fellow at University College London, and a visiting researcher at IBM Watson Research Center and BBN Technologies. He is an area editor of ACM SIGEnergy Energy Informatics Review and an associate editor of IEEE Systems Journal. He is on the program committees of several ACM conferences in smart energy systems and smart cities, such as ACM e-Energy, ACM BuildSys, and ACM MobiHoc. He was a TPC chair of ACM e-Energy 2018, and a guest editor for IEEE Journal on Selected Areas in Communications, IEEE Journal of Internet of Things, and IEEE Transactions on Sustainable Computing. He received a Best Paper Award at ACM e-Energy 2021, a Best Paper Runner-up Award at ACM BuildSys 2018 and has been selected as a number of Best Paper finalists.

Introduction

1

1.1 Background

Cloud computing is a widely popular paradigm nowadays for providing computing, data storage, and digital content services to end-users and enterprises via remote third-party computational resources and infrastructure. Unlike traditional computing, cloud computing users are charged according to their actual usage, without worrying about the sunk costs (e.g., equipment, maintenance and upgrade costs). Cloud computing service providers offer their cloud computing infrastructure, platforms and applications hosted in specialized facilities, as known as *datacenters*.

Because of its benefits and convenience, cloud computing is projected to have a substantial growth of demand in the near future. In addition to traditional data retrieval and query tasks, there are a variety of novel applications that can be hosted in cloud computing, including machine learning, virtual reality, on-demand video streaming, and cloud-based gaming. Along with the accelerating demand for cloud computing, there comes a significant challenge—the energy consumption of datacenters has been skyrocketing. In 2018, datacenters world-wide consumed over 200 terawatt hours of electricity (that is more than 1% of global electricity demand). Such energy consumption is almost equivalent to the total energy consumption of several countries. For example, the energy consumption of datacenters is more than the national energy consumption of Spain [1]. In the US, datacenters alone use more than 90 billion kilowatt-hours of electricity a year, requiring more than 34 large (500-megawatt) power plants. Meanwhile, the energy-related costs are eclipsing the other costs in hardware and facilities of datacenters [2], and are projected to grow around 12% annually [3]. Note that datacenters currently contribute around 15% to the overall carbon emissions of the ICT sector. The overall ICT's carbon footprint is on a similar level with the aviation industry's emissions. However, some predicted that the electricity use by ICT could exceed 20% of the global total by 2030, where a majority of the increase comes

M. Chen and S. C.-K. Chau, *Online Capacity Provisioning for Energy-Efficient Datacenters*, Synthesis Lectures on Learning, Networks, and Algorithms, https://doi.org/10.1007/978-3-031-11549-3_1

from data transmissions in network infrastructure as well as computational processing in datacenters [4].

The accelerating energy consumption in datacenters is a worrying trend. On one hand, we are witnessing a flattening curve of Moore's law. The future growth of microprocessor power has been stagnant. Hence, there will be limited reduction of energy consumption by increasing the computing power in microprocessors per se. On the other hand, the demand for cloud computing has been multiplying. There are substantial demands in cloud computing for machine learning, big data mining, and notoriously computationally intensive cryptocurrency (e.g., Bitcoin) mining. These applications will further exacerbate the energy consumption of datacenters. As a prominent trend for future cloud computing, the explosion of AI applications with Internet-connected devices (as known as AIoT—Artificial Intelligence of Things) will lead to an exponential growth for cloud computing. The number of Internet-connected devices is projected to exceed 20 billion devices by 2020. These devices will require the sophisticated support of real-time artificial intelligence running on cloud computing platforms. Given the fact of current 10 billion Internet-connected devices, doubling that to 20 billion poses a massive increase to the workload of datacenters, which will consequently lead to a significant hike in the electricity consumption in near future. Therefore, mitigating the energy consumption of datacenters, as well as the entailing impacts to the society, is a primary concern of this book.

To formally study the energy consumption of datacenters, let us first define a suitable metric. The energy consumption in a datacenter is usually a product of the so-called *power usage effectiveness* (PUE) and the energy consumption of the computing servers. Specifically, PUE is defined as the ratio between the amount of power entering a datacenter and the power used to run its essential computing facilities. Hence, when PUE is closer to one, the better energy utilization of a datacenter becomes.

There have been several approaches to improve the PUE of datacenters, which can be classified broadly as follows:

- *Endogenous approaches*: These approaches apply to the operations within the datacenters. For example, stripping down the computer servers to the essential functionalities to eliminate non-essential energy consumption. Reducing the air conditioning by open air cooling. One of the most viable and convenient approaches is to use "dynamic provisioning techniques" that reduce energy consumption by turning off unnecessary servers dynamically.
- *Exogenous approaches*: These approaches apply to the external operations of datacenters. For example, deploying on-site renewable energy supplies. Migrating the computational loads by geographical load balancing using demand shifting across multiple datacenters. Also, capturing the wasted heat of datacenters to be reused, for example, as a source for residential heating.

As a result, some so-called hyperscale datacenters (which are specially designed scalable and efficient datacenters) have been established by cloud computing intensive corporations like Amazon, Google, Facebook. A significant energy saving is expected to be made by hyperscale datacenters in terms of PUE. Conventional datacenters have a typical PUE of around 2.0, whereas hyperscale datacenters have a targeted PUE around 1.2. It is reported that Google can attain a PUE of 1.12 on average for its datacenters [4].

In the research literature, there have been numerous studies devoted to topics of energy efficiency of datacenters. For example, recent work has explored electricity price fluctuation in time and geographically balancing computing load across cloud datacenters to reduce the electricity costs; see e.g., [5–8] and the references therein. There were other studies focusing on optimizing cooling [9, 10] and power management [11].

1.2   Dynamic Provisioning in Datacenters

As a basis for effective energy consumption mitigation, it is crucial to mention that the real-world statistics of energy consumption in datacenters reveal three remarkable observations to reduce energy consumption [12–17]. First, the workload in a datacenter often fluctuates significantly on the timescale of hours or days, giving rise to a large "peak-to-mean" ratio. Second, datacenters today often provision for far more than the observed peak to accommodate both the predictable workload and the unpredictable flash crowds. Such static over-provisioning results in low average utilization for most servers. Third, a lightly-utilized or idle server consumes more than 60% of its peak power. These remarkable observations imply that a large portion of the energy consumed by servers goes into powering nearly-idle servers, which can be best saved by turning off servers during the off-peak periods [18].

These observations lead to several useful techniques of optimizing the energy efficiency of computer servers. One promising technique exploiting the above insights is *dynamic provisioning*, which turns on only a minimum number of servers to meet the current demand and dispatches the load among the running servers to meet Service Level Agreements (SLA) in real time, making the datacenter "power-proportional". This can be enabled by virtualization, which is the fundamental technology that allows cloud computing to exist. Virtualization refers to the practice of creating a virtual (rather than physical) version of computing resources, including virtual computer hardware platforms, storage facilities, and network connectivity, with a logical separation from other computing resources that are utilized by different applications or processes.

There has been a significant amount of effort in developing such techniques, initiated by the pioneering work [12, 13] a decade ago. Among them, one line of work [14, 15, 18] examines the practical feasibility and advantage of dynamic provisioning using real-world traces, suggesting that a substantial gain is indeed possible in practice. Another line of work [12, 14, 19] focuses on developing algorithms by utilizing various tools from queuing theory, control theory and machine learning to provide insights that can lead to effective solutions.

These extant studies address a number of solutions that deliver favorable performance, which are supported by theoretical analysis and/or practical evaluations. Readers may refer to [20] for a recent survey.

However, dynamic provisioning incurs additional switching costs and the servers often have inter-dependent properties. These issues make the decisions of dynamic provisioning a non-trivial problem. Typically, it is more effective to predict the future workload (e.g., using model fitting to forecast future workload from historical data [14]) and utilize the predicted information to achieve better performance in dynamic provisioning. This naturally leads to the following fundamental questions:

1. Can we design effective online dynamic provisioning algorithms that require zero future workload information, and yet still achieve close-to-optimal performance?
2. Can we characterize the benefits of utilizing the future workload information, and then improve the design of online algorithms with predictions in dynamic provisioning?

The answers to these questions critically provide a fundamental understanding on how much performance gain one can achieve by exploiting the future workload information in dynamic provisioning, and how to design effective provisioning algorithms for energy-efficient datacenters. The rest of this book is devoted to establishing the results for addressing these questions.

1.3   Related Studies

There has been an extensive body of related work on datacenter energy efficiency, some of which focused on various approaches of energy consumption mitigation from the perspectives of a single processor, an individual datacenter, and about multiple geographically separated datacenters.

We follow the convention of competitive analysis in the literature [21]. The performance of an online algorithm A is often measured by its *competitive ratio*—namely, the maximum ratio of the cost of the solution found by A to the cost of the optimal offline solution that is computed with perfect future knowledge, over all possible problem instances. Competitive analysis, which is an approach from a worst-case perspective, allows us to assess the robust performance guarantee for an online algorithm for arbitrary inputs and arbitrary parameter settings. See Chap. 2 for more details about online algorithms and competitive analysis.

The theoretical studies of energy efficiency in a single speed-scaling processor started by [22]. The authors proposed an offline algorithm that finds a minimum-energy schedule for any set of jobs under the assumption that the unit time power consumption P is a convex function of processor speed s. They also studied two simple online heuristics (Optimal Available and Average Rate) and showed Average Rate has a constant competitive ratio for the case $P(s) = s^\alpha, \alpha \geq 2$. The study [23] considered online dynamic frequency scaling algorithms to minimize the energy used by a server subject to the constraint that every job finishes by its deadline. The authors assumed that the power required to run at frequency f is $P(f) = f^\alpha$ and showed Optimal Available has a competitive ratio upper bounded by α^α. The study [23] also proposed a new online algorithm with competitive ratio $2[\alpha/(\alpha-1)]^\alpha e^\alpha$. The new algorithm is better then Optimal Available for large α in the light of competitive ratio. Stochastic analysis has also been applied to the problem of minimizing power consumption through speed scaling [24]. In their paper [24], the authors optimally scaled speed to balance mean response time and mean energy consumption under processor sharing scheduling. In their paper [25], the authors minimized the average response time of jobs, i.e., the time between their arrival and their completion of service, given the energy budget constraints.

For datacenters with homogeneous servers, the study [26] minimized the Energy-Response time Product (ERP) metric for a datacenter with single application servers which can only run on one frequency but can transition to many *sleep* states while server is idle. For a stationary demand pattern, [26] proved that there exists a very small, natural class of policies that always contains the optimal policy for a single server. For time-varying demand patterns, [26] proposed a simple, traffic-oblivious auto-scaling policy, DELAYEDOFF, to minimizing ERP, and the paper proved that as the average workload ρ goes to infinity, DELAYEDOFF will achieve optimal ERP asymptotically if server can transition from the *off* state to the *on* state instantly. In their paper [27], the authors proposed both offline and online algorithms to minimize the cost of a datacenter with single application environment. The paper proved that the online algorithm is 3-competitive, i.e., its cost is at most 3 times of that of the optimal offline solution. Reference [19, 28] study the problem of minimizing the power cost for a datacenter by combining virtualization mechanisms and DVFS (Voltage and frequency scaling). They proposed optimization models and carried out simulation to evaluate their own models. For datacenters with heterogeneous servers, the study [12] designed an architecture to manage resource for Internet hosting centers through adaptive resource provisioning. The main objective of resource management is to make the hosting centers more energy-efficient. Other papers [29–31] also reduced the energy consumption for heterogeneous datacenters.

Apart from the consideration of an individual datacenter, there are also studies that aim to improve the overall energy-efficiency for multiple geographically separated datacenters [5, 29, 32]. The study [32] minimized the total energy cost for multiple Internet datacenters with location diversity and time diversity of electricity price. They proposed a solution to the constrained mixed-integer programming problem they studied in the paper. They carried out experiments using real data to show that energy cost can be considerably reduced. Reference [29] studied online algorithm "receding horizon control" (RHC) for geographical load balancing and showed that RHC performs well for homogeneous servers. They also presented some variants of RHC with performance guarantees in the face of heterogeneity.

In the area of online algorithm design, there are a number studies that utilize certain information of inputs to achieve better competitive ratio [33–35]. The study [33] assumed that the input (time of skiing) of ski-rental problem is exponentially distributed. Under this assumption, they studied their problem using average-case competitive analysis and proposed optimal online strategy. In their paper [34], the authors used a semi-stochastic model rather than a fully stochastic model to handle input uncertainty in online optimization problems. Specifically, they explored the upper and lower bounds on the amount of stochastic information(online algorithm asks queries to obtain stochastic information about input and each query is confined to following queries: the algorithm gives a value $0 < s < 1$, and then the input gives a value l such that $\int_0^l p(t)\, dt = s$, where $p(x)$ is the real probability distribution of input. More stochastic information means more queries) required by a deterministic algorithm for the ski-rental problem to achieve a desired competitive ratio. In their paper [35], the authors studied online TSP (Traveling Salesman Problem) and TRP (Traveling Repairman Problem). They proposed online algorithms which can utilize the future information (which they called advanced information) to improve the competitive ratios achieved by previous work for the two problems.

In this book, our objective is to reduce the energy cost of a datacenter. The problem studied in this book is similar to that studied in [27]. The difference is that we optimize a linear cost function over integer variables, while Lin et al. in [27] minimize a convex cost function over fractional variables (by relaxing the integer constraints). Note that this book and [27] obtain different online algorithms with different competitive ratios for the two different formulations, respectively. In our formulation, we show that the competitive ratios of our algorithms can be significantly improved by exploiting the look-ahead future information. We believe that looking-ahead provides a valuable degree of freedom in designing "future-aware" online algorithms with desirable competitive ratios. Comparing with [33, 34], we utilize the future workload information in a look-ahead window in our online algorithms, and we study the cases where the future information in the window can be predicted accurately (in both analysis and experiments), or with prediction error (in experiments). This is because the future workload information to a certain extent can be accurately predicted in datacenters [14, 36]. This is also the reason that we do not assume that the future workload follows a certain (partially) known probability distribution in this book.

1.4   Summary of Contributions

In this book, we explore the answers to the fundamental questions about online dynamic provisioning in datacenters, and present the following contributions:

1. We consider a practical scenario where a running server consumes a fixed amount of energy per unit time.We reveal that the dynamic provisioning problem has an elegant structure that allows us to solve it in a "divide-and-conquer" manner. This insight leads to a full characterization of the optimal solution, achieved by a centralized procedure.
2. We show that the optimal solution can also be attained by a simple last-empty-server-first job-dispatching strategy and each server independently solving a classic ski-rental

problem. We build upon this architectural insight to design two decentralized online algorithms. The first is deterministic with competitive ratio $2 - \alpha_s$, where $0 \leq \alpha_s \leq 1$ is the normalized size of a look-ahead window in which future workload information is available. The second is randomized with competitive ratio $\frac{e}{e-1+\alpha_s}$. We prove that $2 - \alpha_s$ and $\frac{e}{e-1+\alpha_s}$ are the best competitive ratios for deterministic and randomized online algorithms under last-empty-server-first job-dispatching strategy.

3. Our results lead to a fundamental observation: under the cost model that a running server consumes a fixed amount of energy per unit time, future workload information beyond the full-size look-ahead window will not improve the dynamic provisioning performance. The size of the full-size look-ahead window is determined by the wear-and-tear cost and the unit-time energy cost of one server. We also believe utilizing future input information is a new and important design degree freedom for online algorithms.

4. We also extend the algorithms to the case where servers take a setup time T_s to be turned on and workload $a(t)$ satisfies $a(\tau) \leq (1 + \gamma)a(t)$ for all $\tau \in [t, t + T_s]$, achieving competitive ratios upper bounded by $(2 - \alpha)(1 + \gamma) + 2\gamma$ and $\frac{e}{e-1+\alpha}(1 + \gamma) + 2\gamma$.

5. Our algorithms are decentralized and efficient to implement. We demonstrated that up to 71% of energy saving can be achieved by our algorithms in a case study using real-world traces. We also compare their performance with other state-of-the-art solutions.

1.5 Organization of the Book

The rest of the book is organized as follows.

- Chapter 2 introduces the preliminaries of online algorithms and competitive analysis.
- Chapter 3 formulated the problem of datacenter capacity provisioning.
- Chapter 4 considers a simplified case with only a single server. We present two online algorithms with proven competitive ratios.
- Chapter 5 generalizes the ideas of the single-server case to the multi-server case. We present two general online algorithms with proven competitive ratios.
- Chapter 6 presents some numerical experiments to corroborate the performance of our algorithms using real-word data.
- Chapter 7 concludes this book with a discussion on future work.

References

1. Country Electricity Ranking, Spain Energy Consumption, http://www.nationmaster.com/country/sp-spain/ene-energy
2. L. Barroso. The price of performance. *ACM Queue*, 3(7):48–53, 2005.
3. U.S. Environmental Protection Agency. Epa report on server and data center energy efficiency. *ENERGY STAR Program*, 2007.

4. N. Jones. How to stop data centres from gobbling up the world's electricity. *Nature*, September 2018.
5. Z. Liu, M. Lin, A. Wierman, S. Low, and L. Andrew. Greening geographical load balancing. In *Proc. ACM SIGMETRICS*, pages 233–244, 2011.
6. P. Wendell, J. Jiang, M. Freedman, and J. Rexford. Donar: decentralized server selection for cloud services. In *Proc. ACM SIGCOMM*, volume 40, pages 231–242, 2010.
7. A. Qureshi, R. Weber, H. Balakrishnan, J. Guttag, and B. Maggs. Cutting the electric bill for internet-scale systems. In *Proc. ACM SIGCOMM*, pages 123–134, 2009.
8. R. Urgaonkar, B. Urgaonkar, M. Neely, and A. Sivasubramaniam. Optimal power cost management using stored energy in data centers. In *Proc. ACM SIGMETRICS*, pages 221–232, 2011.
9. N. Rasmussen. Electrical efficiency modeling of data centers. *Technical Report White Paper*, 113.
10. R. Sharma, C. Bash, C. Patel, R. Friedrich, and J. Chase. Balance of power: Dynamic thermal management for internet data centers. *IEEE Internet Computing*, 2005.
11. R. Raghavendra, P. Ranganathan, V. Talwar, Z. Wang, and X. Zhu. No power struggles: Coordinated multi-level power management for the data center. In *ACM SIGARCH Computer Architecture News*, volume 36, pages 48–59, 2008.
12. J. Chase, D. Anderson, P. Thakar, A. Vahdat, and R. Doyle. Managing energy and server resources in hosting centers. In *Proc. ACM SOSP*, 2001.
13. E. Pinheiro, R. Bianchini, E. Carrera, and T. Heath. Load balancing and unbalancing for power and performance in cluster-based systems. In *Workshop on Compilers and Operating Systems for Low Power*, 2001.
14. G. Chen, W. He, J. Liu, S. Nath, L. Rigas, L. Xiao, and F. Zhao. Energy-aware server provisioning and load dispatching for connection-intensive internet services. In *Proc. USENIX NSDI*, 2008.
15. A. Krioukov, P. Mohan, S. Alspaugh, L. Keys, D. Culler, and R. Katz. Napsac: design and implementation of a power-proportional web cluster. *ACM SIGCOMM Computer Communication Review*, 41(1):102–108, 2011.
16. X. Fan, W. Weber, and L. Barroso. Power provisioning for a warehouse-sized computer. In *Proc. the 34th annual international symposium on Computer architecture*, 2007.
17. L. Barroso and U. Holzle. The case for energy-proportional computing. *IEEE Computer*, 40(12):33–37, 2007.
18. D. Meisner, B. Gold, and T. Wenisch. Powernap: eliminating server idle power. *ACM SIGPLAN Notices*, 2009.
19. H. Qian and D. Medhi. Server operational cost optimization for cloud computing service providers over a time horizon. In *Proceedings of the 11th USENIX conference on Hot topics in management of internet, cloud, and enterprise networks and services*, pages 4–4, 2011.
20. A. Beloglazov, R. Buyya, Y. C. Lee, and A. Zomaya. A taxonomy and survey of energy-efficient data centers and cloud computing systems. *Advances in Computers, M. Zelkowitz(ed.), vol. 82, pp. 47-111*, 2011.
21. A. Borodin and R. El-Yaniv. *Online Computation and Competitive Analysis*. Cambridge University Press, 2005.
22. F. Yao, A. Demers, and S. Shenker. A scheduling model for reduced cpu energy. In *Foundations of Computer Science, 1995. Proceedings., 36th Annual Symposium on*, pages 374–382. IEEE, 1995.
23. N. Bansal, T. Kimbrel, and K. Pruhs. Dynamic speed scaling to manage energy and temperature. 2004.
24. A. Wierman, L. Andrew, and A. Tang. Power-aware speed scaling in processor sharing systems. In *INFOCOM 2009, IEEE*, pages 2007–2015. IEEE, 2009.

25. K. Pruhs, P. Uthaisombut, and G. Woeginger. Getting the best response for your erg. *Algorithm Theory-SWAT 2004*, pages 14–25, 2004.
26. A. Gandhi, V. Gupta, M. Harchol-Balter, and M. Kozuch. Optimality analysis of energy-performance trade-off for server farm management. *Performance Evaluation*, 2010.
27. M. Lin, A. Wierman, L. Andrew, and E. Thereska. Dynamic right-sizing for power-proportional data centers. In *Proc. IEEE INFOCOM*, pages 1098–1106, 2011.
28. V. Petrucci, O. Loques, and D. Mossé. Dynamic optimization of power and performance for virtualized server clusters. In *Proceedings of the 2010 ACM Symposium on Applied Computing*, pages 263–264, 2010.
29. M. Lin, Z. Liu, A. Wierman, and L. L. H. Andrew. Online algorithms for geographical load balancing. In *Proc. Int. Green Computing Conf.*, 2012.
30. R. Nathuji, C. Isci, and E. Gorbatov. Exploiting platform heterogeneity for power efficient data centers. In *Autonomic Computing, 2007. ICAC'07. Fourth International Conference on*, pages 5–5. IEEE, 2007.
31. T. Heath, B. Diniz, E. Carrera, W. Meira Jr, and R. Bianchini. Energy conservation in heterogeneous server clusters. In *Proceedings of the tenth ACM SIGPLAN symposium on Principles and practice of parallel programming*, pages 186–195. ACM, 2005.
32. L. Rao, X. Liu, L. Xie, and W. Liu. Minimizing electricity cost: Optimization of distributed internet data centers in a multi-electricitymarket environment. *Proc. IEEE INFOCOM*, 2010.
33. H. Fujiwara and K. Iwama. Average-case competitive analyses for ski-rental problems. *Algorithms and Computation*, pages 157–189, 2002.
34. A. Mądry and D. Panigrahi. The semi-stochastic ski-rental problem.
35. P. Jaillet and M. Wagner. Online routing problems: Value of advanced information as improved competitive ratios. *Transportation Science*, pages 200–210, 2006.
36. P. Bodík, R. Griffith, C. Sutton, A. Fox, M. Jordan, and D. Patterson. Statistical machine learning makes automatic control practical for internet datacenters. In *Proceedings of the 2009 conference on Hot topics in cloud computing*, 2009.

Preliminaries of Online Algorithms and Competitive Analysis

2

This chapter presents the basic ideas of online algorithms and competitive analysis. It aims to establish the foundation for the online algorithms for datacenter capacity provisioning in the subsequent chapters.

2.1 Introduction

The more you know the better decisions you can make, particularly when making decisions regarding the future. If you know all the future events in advance, then you will certainly be able to prepare for all the anticipated contingencies in the best possible manners. Undoubtedly, the knowledge of future events can have a critical impact on the present decisions in many problems. However, in practice we often face the situations where not all information necessary to the current decisions is available before the moment at which the decisions are determined. In this case, the decision-making processes will require non-trivial techniques and smart strategies to mitigate the adverse effect of incomplete knowledge of the future events.

The decision-making problems with incomplete knowledge of future events often occur in our daily lives. For decades, computer scientists and operations researchers have been studying sequential decision-making processes with a sequence of gradually revealed events, and analyzing the best possible strategies considering incomplete knowledge of future events. For example, see the comprehensive textbook [1] and the references therein. These strategies are commonly known as *online algorithms*. The analysis of online algorithms considering the worst-case impacts of incomplete knowledge of future events is called *competitive analysis*. Competitive analysis is a natural benchmark for online algorithms. This chapter presents some classical problems of online algorithms and provides competitive analysis of these online algorithms.

© The Author(s), under exclusive license to Springer Nature Switzerland AG 2022
M. Chen and S. C.-K. Chau, *Online Capacity Provisioning for Energy-Efficient Datacenters*, Synthesis Lectures on Learning, Networks, and Algorithms, https://doi.org/10.1007/978-3-031-11549-3_2

2.2   Model of Online Optimization

We define a formal model of online optimization. We are interested in the class of *sequential decision-making* optimization problems, whereby a sequence of decisions for certain actions that optimize a certain objective are determined over time according to the observations and environmental parameters that are acquired gradually in a sequential manner. As we will see in the subsequent chapters, a key example is a datacenter scenario with computing demands and available supplies of resources arriving in an ad hoc manner. In a datacenter, the computing demands may be varied dynamically, and the supplies of resources may not be sufficient momentarily. The goal of this book is to shed light on a desirable decision-making process that balances the demands and supplies in an intelligent manner, in order to optimize a performance objective of the operations in a datacenter.

2.2.1   Arrival Models

Formally, we model the inputs as a finite sequence of arrivals of events (e.g., observations, system inputs, and environmental parameters). There are two formal models describing the inputs:

1. *Time Stamp Model*, in which each arrival of an event is associated with a definite time stamp. For example, we can denote a sequence of timestamps by $(t_1, t_2, ...)$, where each arrival of an event is denoted by a function of a timestamp, $a(t)$.
2. *Sequence Model*, in which the arrivals of events are represented by an abstract sequence assuming certain pre-defined time intervals. For example, we can denote a sequence of arrivals by a sequence $(a_1, a_2, ...)$.

There are subtle differences between the two models. From the perspective of a decision-maker, the main difference is that the decision-maker in the sequence model will implicitly execute the action of each decision before the next arrival of the event, whereas the execution time of action is explicitly recorded in the time stamp model. Such a difference may have an impact on the overall objective value. For example, the latency of action may incur a certain penalty, so that there is a difference between the action executed immediately after an observed event, and that before the next observed event. But we note that it is also possible to define an abstract sequence model based on a time stamp model by incorporating the effect of delayed actions as a cost in the objective value. Sometimes, the sequence model is interpreted as a discrete-time model. Note that throughout this book, we will assume the sequence model in our study of datacenters, unless stated otherwise.

2.2.2 Offline Optimization Versus Online Optimization

Before we discuss decision-making without complete knowledge of the future, we consider the ideal setting with complete knowledge of future events, which is known as *offline optimization*. In such a setting, the planning process can fully utilize the future knowledge to determine the best possible decisions. Sometimes, offline optimization is studied as a hypothetical analysis assuming the best possible a-priori knowledge (i.e., with an oracle).

In reality, it is unlikely that all information necessary to solve a problem instance is available beforehand. For example, the demands may arrive earlier than any supplies become available. By the time of actions, certain decisions anticipating future contingencies are required. Thus, this would require a different way of decision-making processes, which is known as *online optimization*. An optimization algorithm is called online if it makes a decision at the immediate moment when a new piece of information is revealed to request an action, rather than making all the actions after the full information is revealed at the end. For example, when a demand arrives, the system needs to immediately decide how it can be served without waiting for further demands. In other words, an online algorithm is akin to making real-time decisions in a sequentially timely manner, without waiting for the posterior knowledge to be revealed in future.

Because of the temporal properties, online algorithms differ considerably from traditional (offline) algorithms in the way of how problems are solved. Note that every online optimization problem has a counterpart of offline optimization, which assumes that each decision is made based on an oracle that reveals all the future information in advance. Sometimes, knowing how decisions are made in offline optimization can provide a valuable lesson on the design of proper online optimization. Therefore, analyzing offline optimization on a problem is a cornerstone for understanding its counterpart of online optimization.

2.2.3 Competitive Analysis

The first question we would ask is what makes a good online algorithm. While online algorithms have obvious disadvantages over offline algorithms because they cannot see the future information, it is important to understand how such a limitation will restrict the way of solving a problem, and to gain insights on the effectiveness of online optimization. To address this question, we need a general metric to characterize the effectiveness of online algorithms.

An important metric of effectiveness for online algorithms is based on *competitive analysis*. The basic idea of competitive analysis is to benchmark an online algorithm against a corresponding optimal offline algorithm in the worst case of inputs. Competitive analysis can characterize the maximum impact of incomplete future information to decision-making processes. The worst-case ratio between the cost of an online algorithm and that of a corresponding optimal offline algorithm on the same inputs is known as *competitive ratio*.

Competitive ratio is a natural benchmark for the comparison between online algorithms and offline algorithms.

There are both advantages and disadvantages of benchmarking against a corresponding optimal offline algorithm. Prediction is a common feature in temporal decision-making processes. An optimal offline algorithm can be regarded as the best possible prediction-based algorithm, because it has the perfect future knowledge. Naturally, a good online algorithm should perform as close as possible to an optimal offline algorithm. If we can bound the gap between an online algorithm against a corresponding optimal offline algorithm, namely, by finding the worst case of inputs, then it will provide assurance to other (better than the worst-case) inputs. Moreover, there are more complicated prediction-based algorithms using machine learning. It is difficult to benchmark against those algorithms, whereas an optimal offline algorithm provides a simple case for comparison, which also provides a lower bound for the gap between online algorithms and other prediction-based algorithms.

However, on the other hand, competitive analysis is sometimes too pessimistic for online algorithms, because the worst-case inputs will seldom appear in practice. Furthermore, the empirical ratio between an online algorithm and a corresponding optimal offline solution, observed in the average case, can be much better than that of competitive analysis. Although it is possible to compare an online algorithm and a corresponding optimal offline solution based on average-case inputs, it is not straightforward to have a proper definition of average-case inputs. Average-case analysis requires case-by-case justifications for specific applications.

2.2.4 Definitions

In the following, we formally define the notions of an online algorithm and its competitive ratio. Let the input sequence be $a = (a_1, a_2, ..., a_t, ...)$ and the corresponding sequence of decisions be $x = (x_1, x_2, ..., x_t, ...)$, where a_t and x_t are the current input and decision at time t, respectively. Let the cost function with respect to the sequence of decisions x be $\text{Cost}(x)$.

Definition 2.1 (*Deterministic Online Algorithm*) A deterministic online algorithm $\mathcal{A}$ can determine a unique decision x_t at time t, given the inputs revealed at time t or before. Namely, we can write $\mathcal{A}(a_1, ..., a_t) = x_t$. We also write $\mathcal{A}[a] = x$ as the whole sequence of decisions output by $\mathcal{A}$.

To analyze the effectiveness of a deterministic online algorithm $\mathcal{A}$, we adopt an *adversary model* in competitive analysis, whereby an adversary who controls the input sequence knows the algorithmic code of $\mathcal{A}$. The goal of the adversary is to generate the worst-case inputs to inflict the maximum cost on $\mathcal{A}$, as compared with an optimal offline solution (by an oracle).

Definition 2.2 (*Competitive Ratio*) A deterministic online algorithm $\mathcal{A}$ is called c-competitive (or has a competitive ratio c), if the objective function value of the solution produced by $\mathcal{A}$ on any input sequence a generated by an adversary is at most c times that of an optimal offline algorithm on the same input. Here, an optimal offline solution that has complete knowledge about the whole input sequence is denoted by Opt[a]. Formally, the competitive ratio of $\mathcal{A}$, denoted by CR[$\mathcal{A}$], is defined as follows:

$$\text{CR}[\mathcal{A}] \triangleq \max_{a} \frac{\text{Cost}(\mathcal{A}[a])}{\text{Cost}(\text{Opt}[a])}$$

In the above definition, the competitive ratio CR[$\mathcal{A}$] is taken over any input sequence a, which measures the worst-case scenario.

Definition 2.3 (*Randomized Online Algorithm*) A randomized online algorithm $\mathcal{R}$ is a probability distribution over a set of deterministic online algorithms. By making decisions probabilistically, $\mathcal{R}$ produces an ensemble of decisions based on the same input sequence a.

We remark that a randomized algorithm does not necessarily entail random decisions at each execution of the algorithm. Alternatively, when we consider an ensemble of a large number of executions at different times, each execution can be given a deterministic decision rule drawn from a probabilistic ensemble of decision rules. The expected cost of randomized algorithm $\mathcal{R}$ is denoted by $\mathbb{E}[\text{Cost}(a, \mathcal{R}[a])]$, where $\mathbb{E}[\cdot]$ is the expectation over all random decisions of $\mathcal{R}$.

We assume that $\mathcal{R}$ has to generate all random bits in advance before making any decisions. Note that this does not limit the applicability of our analysis, because the sequence of random bits from a pseudo random generator can be determined by setting the seed. In the competitive analysis of a randomized online algorithm $\mathcal{R}$, an adversary can be modeled with different levels of knowledge. (1) *Oblivious Adversary Model*: In the basic setting, the adversary knows the algorithmic code but does not know the random bits used by $\mathcal{R}$. Equivalently, the adversary needs to generate the whole input sequence before observing the random decisions made by $\mathcal{R}$. (2) *Adaptive Adversary Model*: In a more sophisticated setting, the adversary knows both knows the algorithmic code and random bits used by $\mathcal{R}$. The oblivious adversary model is a more natural model to analyze the effectiveness of randomized online algorithms in practice against nature (rather than a malicious attacker). In the following, we will assume the oblivious adversary model.

Definition 2.4 (*Expected Competitive Ratio*) If a randomized online algorithm $\mathcal{R}$ is considered, the competitive ratio is taken in expectation with respect to the probability distribution of random decisions of $\mathcal{R}$. The expected competitive ratio of randomized algorithm $\mathcal{R}$ with respect to an oblivious adversary is defined as follows:

$$CR[\mathcal{R}] \triangleq \max_a \frac{\mathbb{E}[Cost(\mathcal{R}[a])]}{Cost(Opt[a])}$$

In the above definition, the competitive ratio CR[$\mathcal{A}$] is considered in the expected cost of randomized algorithm $\mathcal{R}$ over any input sequence a, which measures the worst-case scenario in expectation.

2.3 Ski-Rental Problem

In this section, we will tackle a classical online optimization problem, called *ski-rental problem*, which is also known as buy-or-rental problem. Competitive analysis will be then presented to analyze the effectiveness of the related online algorithms.

In this problem, imagine you are at a skiing resort. Suppose that you are new to skiing and don't own any skiing gear. As a result, you should either rent skiing gear for a couple of days, or buy them once and use them for a longer period. However, you do not know how long you will need them (for example, you may not know how long you will enjoy skiing). *Should you rent or buy your skiing gear, without knowing the number of days you will ski in the future?* We remark that this is not just an artificial problem. Indeed, we face numerous similar practical situations in real life, making decisions in an uncertain environment between a more flexible but costly option (e.g., renting) and another cheaper but more committed option (i.e., buying). For example, shall you take a taxi (which is faster but subject to variable traffic) or a train (which is slower but more predictable)? Shall you take out a loan at a market-variable interest rate or a fixed interest rate? These problems can be reformulated as some variants of the classical ski-rental problem.

Formally, we formulate the ski-rental problem as follows. We consider discrete-time settings, with daily intervals. Assume that renting skiing gear costs \$1 per day and buying one costs \$B. It is reasonable that $B > 1$. The decision-maker has to decide in an online fashion whether should continue renting or buy skiing gear on each day to minimize the cost, only knowing whether he still skis on the present day. One the other hand, the adversary, knowing the decision-maker's decision strategy, will select the total number of days of skiing (denoted by k) to maximize the cost. In competitive analysis, we aim to devise a decision strategy for the decision-maker to minimize the worst-case ratio of the cost decided by an adversary over the optimal cost of an oracle.

2.3.1 Optimal Offline Algorithm

In the offline setting, the parameters are known in advance. Namely, we know the exact number of days of skiing, k. Hence, with the knowledge of k, one can design the optimal algorithm for making the decisions in the ski-rental problem. If $k \leq B$, then it is better off to rent the skiing gears for the rest of the days. On the other hand, if $k \geq B$, then it is better off to buy the skiing gears on the first day. Hence, the optimal cost is $\min\{k, B\}$. We next design competitive online algorithms compared with the optimal offline algorithm.

2.3.2 Deterministic Threshold-Based Online Algorithm

Without knowing the number of days of skiing, how would you decide to buy or rent skiing gear? Let us think about the possible decision strategies. On each day, the decision is only dependent on two inputs: (1) whether skiing is still needed on the present day, and (2) how many days the decision-maker has been skiing so far. Hence, any decision strategy will be essentially a function that maps the inputs to the decision on that day.

A simple decision strategy will be first choosing a threshold θ, then renting for up to $\theta - 1$ days and next buying the skiing gear on the θ-th day. This is a simple deterministic online algorithm, which does not depend on the knowledge of k. We define a few notations. Let the input be $a_t = 1$ if still skiing on the t-th day, otherwise, 0. Let decision $x_t = 1$ represent renting, whereas $x_t = 0$ represents buying at t. The pseudo-code of such a deterministic online algorithm is presented in **OnSki**$[\theta]$ (see Algorithm 2.1).

Algorithm 2.1 Deterministic threshold-based online algorithm **OnSki**$[\theta]$

Parameter : Threshold θ
Input : $a_1, ...a_t$
Output : x_t

1: Compute the number of days of skiing known so far: $\hat{k} \leftarrow \sum_{\tau=1}^{t} a_\tau$
2: **if** $\hat{k} < \theta$ **then**
3:    Continue renting: $x_t \leftarrow 1$
4: **else**
5:    Buying: $x_t \leftarrow 0$
6: **end if**
7: Return x_t

We next analyze the competitive ratio of threshold-based **OnSki**$[\theta]$ in Theorem 2.5.

Theorem 2.5 *The competitive ratio of deterministic algorithm* **OnSki**[θ] *is given by*

$$\text{CR}\Big[\textbf{OnSki}[\theta]\Big] = \frac{\theta - 1 + B}{\min\{\theta, B\}}$$

and the minimum competitive ratio of **OnSki**[θ] *is attained when setting* $\theta = B$, *such that*

$$\min_{\theta} \ \text{CR}\Big[\textbf{OnSki}[\theta]\Big] = 2 - \frac{1}{B}$$

Proof If $\theta > k$, then the cost of **OnSki**[θ] is given by

$$\text{Cost}(\textbf{OnSki}[\theta]) = k$$

On the other hand, if $\theta \le k$, then the cost is given by

$$\text{Cost}(\textbf{OnSki}[\theta]) = \theta - 1 + B$$

The adversary aims to maximize the competitive ratio by setting an appropriate value of k. Note that the optimal cost is $\min\{k, B\}$. Given θ, the adversary can inflict the maximum ratio $\frac{\text{Cost}(\textbf{OnSki}[\theta])}{\min\{k, B\}}$ by setting $k = \theta$. Otherwise, setting $k > \theta$ or $k < \theta$ will only obtain a smaller value of $\frac{\text{Cost}(\textbf{OnSki}[\theta])}{\min\{k, B\}}$. Therefore, the competitive ratio of **OnSki**[θ] is

$$\text{CR}\Big[\textbf{OnSki}[\theta]\Big] = \frac{\theta - 1 + B}{\min\{\theta, B\}}$$

Next, we define a function $f(\theta)$:

$$f(\theta) \triangleq \frac{\theta - 1 + B}{\min\{\theta, B\}} = 1 + \frac{\max\{\theta, B\} - 1}{\min\{\theta, B\}}$$

$f(\theta)$ will be minimized at $\theta = B$, which gives the minimum competitive ratio $2 - \frac{1}{B}$. Otherwise, setting $\theta > B$, we obtain

$$f(\theta) = 1 + \frac{\theta - 1}{B} < 2 - \frac{1}{B}$$

Also, setting $\theta < B$, we obtain

$$f(\theta) = 1 + \frac{B - 1}{\theta} < 2 - \frac{1}{\theta} < 2 - \frac{1}{B}$$

This completes the proof. $\square$

2.3.3   Lower Bound for Deterministic Online Algorithms

Although threshold-based **OnSki**[θ] is a plausible online algorithm, the question is whether **OnSki**[θ] is the best deterministic online algorithm for solving ski-rental problem. Theorem 2.6 can provide an affirmative answer.

Theorem 2.6 *There is no deterministic online algorithm for solving ski-rental problems that can achieve a competitive ratio* $< 2 - \frac{1}{\min\{k, B\}}$.

Proof For any online algorithm $\mathcal{A}$, we show that there exists an input sequence a, such that

$$\text{Cost}(\mathcal{A}[a]) \geq 2 \cdot \text{Cost}(\text{Opt}[a]) - 1$$

First, we consider an infinite input sequence $(a_t)_t$, such that $a_t = 1$ for all t. If $\mathcal{A}[a]$ never buys skiing gear on any finite t, then $\text{Cost}(\mathcal{A}[a]) \to \infty$ and the theorem holds.

Otherwise, assume that $\mathcal{A}[a]$ buys skiing gear on the t'-th day. Then, the adversary will set $k = t'$. Hence, $\text{Cost}(\mathcal{A}[a]) = (k - 1) + B = k + B - 1$, whereas $\text{Cost}(\text{Opt}[a]) = \min\{k, B\}$.

Therefore, we obtain

$$\text{Cost}(\mathcal{A}[a]) \geq 2 \cdot \min\{k, B\} - 1 = 2 \times \text{Cost}(\text{Opt}[a]) - 1$$

Hence, $\text{CR}[\mathcal{A}] \geq 2 - \frac{1}{\min\{k, B\}}$ for any online algorithm $\mathcal{A}$.   □

2.3.4   Randomized Threshold-Based Online Algorithm

Instead of fixing one particular threshold in **OnSki**[θ], we can employ a simple randomized online algorithm that first probabilistically choose a threshold $\hat{\theta}$ according to a pre-defined probability distribution $\rho(\hat{\theta})$ and then run $\mathcal{A}_{\hat{\theta}}$. The pseudo-code of such a randomized online algorithm is presented in **ROnSki**[ρ] (see Algorithm 2.2).

Theorem 2.7 *Let the probability distribution* $\rho(\hat{\theta})$ *for* **ROnSki**[ρ] *be*

$$\rho(\hat{\theta}) = \left(\frac{B - 1}{B}\right)^{B - \hat{\theta}} \frac{c}{B},$$

where $\hat{\theta} \in \{0, 1, 2, 3, \ldots\}$ *and we set* $c = \frac{1}{1 - (1 - \frac{1}{B})^B}$. *Then the expected competitive ratio of randomized algorithm* **ROnSki**[ρ] *is* c

Note that $\frac{1}{1 - (1 - \frac{1}{B})^B}$ is an increasing function in B, when $B > 1$. Also, $\lim_{B \to \infty} \frac{1}{1 - (1 - \frac{1}{B})^B} = \frac{e}{e - 1}$.

Algorithm 2.2 Randomized threshold-based online algorithm **ROnSki[ρ]**

Parameter : Probability distribution $\rho(\hat{\theta})$
Input : $a_1, \ldots a_t$
Output : x_t

1: Pick $\hat{\theta}$ with probability $\rho(\hat{\theta})$
2: Set $x_t \leftarrow$ **OnSki[$\hat{\theta}$]**
3: Return x_t

When we compare the competitive ratios of deterministic algorithm **OnSki[θ]** and randomized algorithm **ROnSki[ρ]**, we obtain the difference as

$$2 - \frac{1}{B} - \frac{1}{1 - (1 - \frac{1}{B})^B} \leq 2 - \frac{e}{e-1} \approx 0.418$$

Hence, randomized algorithm **ROnSki[ρ]** can improve the competitive ratio of deterministic algorithms **OnSki[θ]** by around 21%.

In the following, we consider the continuous-time setting such that the number of days is a positive real number, and prove a simpler version of Theorem 2.7.

Recall that $\text{Cost}(\text{Opt}[a]) = \min\{k, B\}$. Given a probability distribution $\rho(\hat{\theta})$ such that $\int_0^B \rho(\hat{\theta})d\hat{\theta} = 1$, the competitive ratio of **ROnSki[ρ]** can be related by

$$c = \frac{1}{x} \cdot \left(\int_0^x (B + \hat{\theta})\rho(\hat{\theta})d\hat{\theta} + x \int_x^B \rho(\hat{\theta})d\hat{\theta} \right)$$

where $x = \min\{k, B\}$, and $0 \leq x \leq B$. We rearrange the equation as follows:

$$\left(\int_0^x (B + \hat{\theta})\rho(\hat{\theta})d\hat{\theta} + x \int_x^B \rho(\hat{\theta})d\hat{\theta} \right) = cx$$

Differentiating the above equation with respect to x using differentiation product rule, we obtain

$$(B + x)\rho(x) + \int_x^B \rho(\hat{\theta})d\hat{\theta} + x(-\rho(x)) = c \tag{2.1}$$

Differentiating again and noting that by continuity we should have $\rho(B) = 0$:

$$\rho(x) + (B + x)p'(x) - \rho(x) - xp'(x) - \rho(x) = 0$$

In other words, $\frac{p'(x)}{\rho(x)} = \frac{1}{B}$, and hence we obtain $\rho(x) = Ke^{\frac{x}{B}}$. Since $\int_0^B \rho(\hat{\theta})d\hat{\theta} = 1$, it follows that $KB(e - 1) = 1$ and $K = \frac{1}{B(e-1)}$. Finally, using Eq. (2.1) and considering $x = 0$

(such that $\mathsf{Cost}(\mathsf{Opt}[a])$ is minimized), we obtain

$$c = B\rho(0) + \int_0^B \rho(\hat{\theta})d\hat{\theta} = \frac{e}{e-1}$$

2.3.5   Lower Bound for Randomized Online Algorithms

The next question is whether randomized threshold-based online algorithm **ROnSki**[ρ] is the best randomized algorithm for solving ski-rental problem, which will be addressed by Theorem 2.8.

Theorem 2.8 *There is no randomized online algorithm that can achieve an expected competitive ratio* $< \dfrac{1}{1-\left(1-\frac{1}{B}\right)^{B+1}}$.

We sketch the proof idea of the above theorem. First, we introduce the idea of Yao's minimax principle. This principle will be useful to provide a lower bound of the competitive ratio of randomized algorithms.

Yao's minimax principle (see [1] for a detailed account) is stated as follows: Given a randomized online algorithm $\mathcal{R}$, the cost of $\mathcal{R}$ can be lower bounded by

$$\max_a \mathbb{E}[\mathsf{Cost}(\mathcal{R}[a])] \geq \min_{\mathcal{A}} \mathbb{E}[\mathsf{Cost}(\mathcal{A}[A])] \tag{2.2}$$

where $\mathcal{A}$ is any deterministic online algorithm and A is any given random input. Note that the later expectation operator is taken with respect to random input A. In Yao's minimax principle, we assume an oblivious adversary model, such that the adversary does not know the random bits used by $\mathcal{R}$.

We next construct a random input A, and show that

$$\mathsf{CR}[\mathcal{R}] \geq \min_A \frac{\mathbb{E}[\mathsf{Cost}(\mathcal{A}[A])]}{\mathbb{E}[\mathsf{Cost}(\mathsf{Opt}[A])]} = \frac{e}{e-1} \tag{2.3}$$

We construct the following random input. Let $P(k \geq y) = \left(1 - \frac{1}{B}\right)^y$ be the probability distribution of k in input A. One can show that $\mathbb{E}[\mathsf{Cost}(\mathsf{Opt}[A])] = B\left(1 - \left(1 - \frac{1}{B}\right)^{B+1}\right)$ and $\mathbb{E}[\mathsf{Cost}(\mathcal{A}(A)] \geq B$ for any deterministic online algorithm $\mathcal{A}$. Hence, by Yao's minimax principle, this shows

$$\mathsf{CR}[\mathcal{R}] \geq \frac{1}{1 - \left(1 - \frac{1}{B}\right)^{B+1}}$$

Note that $\lim_{B \to \infty} \dfrac{1}{1-\left(1-\frac{1}{B}\right)^{B+1}} = \dfrac{e}{e-1}$.

Remarks The ski-rental problem is a fundamental problem, which is often embedded in many practical problems. The above results of the ski-rental problem can be used as a basic component in the design of more advanced online algorithms for other online decision-making problems. In the subsequent chapters, we will apply some online algorithms similar to those of the ski-rental problem to the problem of datacenter capacity provisioning.

2.4   Summary

In this chapter, we presented some preliminaries of online algorithms and competitive analysis. Particularly, we presented the deterministic and randomized online algorithms for tackling the classic ski-rental problem with a theoretical analysis of their competitive ratios. We showed that these online algorithms can attain the best achievable competitive ratios. These algorithms will be extended for the more complex datacenter capacity provisioning problem, as presented in the next chapter.

Reference

1. A. Borodin and R. El-Yaniv. *Online Computation and Competitive Analysis*. Cambridge University Press, 2005.

Modeling and Problem Formulation

3

This chapter presents the formulation and models of the datacenter capacity provisioning problem. We will derive competitive online algorithms with proven competitive ratios to solve the datacenter capacity provisioning problem effectively in the subsequent chapters.

3.1   Settings and Models

First, we adopt the models of datacenters from the extant literature [1–8]. Our models can realistically capture several key features of datacenters, including workload, server energy consumption, cooling and power conditioning systems, and dependence of power grid. But these models are not meant to be completely extensive. Some features are dropped for the sake of brevity. Nonetheless, in the discussion section (Sect. 3.2.3), we will present further extensions of the datacenter model.

We consider a discrete-time model whose time time slot matches the timescale at which the decisions are made. Without loss of generality, we assume that there are totally T time slots, and each has a unit length. We consider a datacenter consisting of a set of homogeneous servers. Without loss of generality, we assume that each server has a unit service capacity, i.e., it can only serve one unit workload per unit time. In practice, a server's service capacity can be determined from the knee of its throughput and response-time curve [9].

The system model in our datacenter capacity provisioning problem consists of three major parts: (1) workload model, (2) datacenter model, and (3) power grid model. The datacenter model is further divided into (1) server energy model, (2) power conditioning system and (3) cooling system model. See Fig. 3.1 for a graphical representation. Next, we describe each of the models in the following sections.

Before presenting the models, we note that some key notations used in the system model are summarized in Table 3.1.

© The Author(s), under exclusive license to Springer Nature Switzerland AG 2022

M. Chen and S. C.-K. Chau, *Online Capacity Provisioning for Energy-Efficient Datacenters*, Synthesis Lectures on Learning, Networks, and Algorithms, https://doi.org/10.1007/978-3-031-11549-3_3

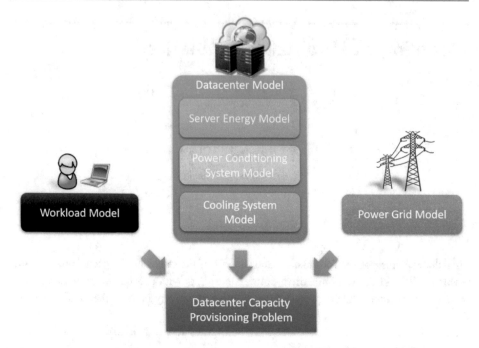

Fig. 3.1   A graphical representation of the system model

3.1.1   Workload Model

The workload model aims to capture the demands for cloud computing services at the datacenter. Similar to extant work [5, 10, 11], we consider a "mice" type of workload for the datacenter where each job has a small transaction size and short duration. The jobs arriving in a time slot will be served in the same slot. The workload can be split among multiple active servers at arbitrary granularity like a fluid. These assumptions model a "request-response" type of workload that characterizes serving web content or hosted application services that entail short but real-time interactions between the user and the server. The workload to be served at time $t \in [1, T]$ is represented by $a(t)$. Note that in this book, we do not make any stochastic assumption about the probability distribution of the workload $a \triangleq \{a(t)\}_{t=1}^{T}$.

3.1.2   Server Energy Model

We consider a hyperscale datacenter, with computing resources that can be provisioned flexibly in a distributed computing environment. As a result, a hyperscale datacenter is able to scale and grow systems quickly, from a few servers to tens of thousands. In such a datacenter,

Table 3.1 Some key notations. Note that we use bold symbols to denote vectors, e.g., $x = \langle x(t) \rangle$. Brackets indicate the corresponding units

Notation	Definition
T	Number of time time slots
$a(t)$	Workload at time t
$x(t)$	Number of active ("on") servers at time t
c_{idle}	Server power at idling state (Watt)
c_{peak}	Server power at fully active state (Watt)
β_s	Switching cost of a server ($)
$f_p(x(t), a(t))$	Server power consumption as a function of $x(t)$ and $a(t)$ at time t (Watt)
$b(t)$	Total power consumption of server at time t
$f_b(b(t))$	Total power consumption of power conditioning system as a function of $b(t)$ (Watt)
$f_c^t(b(t))$	Total power consumption of cooling system as a function of $b(t)$ at time t (Watt)
$p(t)$	Price per unit energy acquired from power grid at time t ($P_{\text{min}} \le p(t) \le P_{\text{max}}$) ($/Wh)
$v(t)$	Units of energy used from power grid at time t (Watt)
$g_t(x(t), a(t))$	Total power consumption as a function of $x(t)$ and $a(t)$ at time t (Watt)
$\text{Cost}(x, v)$	Total operating cost of datacenter as a function of $x(t)$ and $v(t)$ at t ($)
$[\cdot]^+$	Operator $\max(0, \cdot)$

the computing servers can be turned on or off dynamically by system administrators or automatic management software.

The server energy model aims to capture the energy consumption of servers in a hyperscale datacenter. Denote the number of servers "on" (serving or idle) at time t by $x(t)$. We model the aggregate server power consumption by

$$b(t) \triangleq f_s(x(t), a(t)),$$

which is an increasing (i.e., non-decreasing) and convex function of $x(t)$ and $a(t)$. That is, the first and second order partial derivatives in $x(t)$ and $a(t)$ are all non-negative. In addition, to get the workload served in the same slot, we must have $x(t) \ge a(t)$.

This server energy model is rather general and covers many common types as special cases. One example is the commonly adopted standard linear model [12]:

$$f_s\left(x(t), a(t)\right) = c_{\texttt{idle}}x(t) + \left(c_{\texttt{peak}} - c_{\texttt{idle}}\right)a(t),$$

where $c_{\texttt{idle}}$ and $c_{\texttt{peak}}$ are the power of a server in the idling and the fully-utilized states, respectively. Most servers today consume a significant amount of energy even in the idling state. A holy grail in the server design is to make them "power proportional" by reducing $c_{\texttt{idle}}$ to as close to zero as possible [13].

In addition to the power cost, turning a server on entails a switching cost [14], denoted as β_s, which normally includes the amortized service interruption cost, wear-and-tear cost, e.g., component procurement, replacement cost (hard-disks in particular) and risk associated with server switching. We note that the on/off cost is comparable to the energy cost of running a server for several hours [2], and hence, is not negligible.

3.1.3 Power Conditioning System Model

The term power conditioning system refers to a general class of devices that use power electronics technologies to convert electric power from one form to another. The power conditioning systems in a datacenter usually include power distribution units (PDUs) and uninterruptible power supplies (UPSs). PDUs transform the high voltage power distributed throughout the datacenter to voltage levels appropriate for servers. UPSs provides temporary power during outages.

It is important to model the power consumed by the power conditioning and cooling systems, as server, power conditioning, and cooling contribute about 94% of the overall power consumption in a datacenter, and their power consumptions vary drastically with server utilization [15]. We model the power consumption of the system at a given time t as $f_p(b(t))$, an increasing and convex function of the aggregate server power consumption $b(t)$.

This model is general and one example is the quadratic function adopted in a comprehensive study on the datacenter power consumption [15]:

$$f_p(b(t)) = C_1 + L_1 b^2(t),$$

where $C_1 > 0$ and $L_1 > 0$ are constants depending on the additional physical models of PDUs and UPSs. Here, we rely on the models and parameters from the related literature [15].

3.1.4 Cooling System Model

Datacenters dissipate an enormous amount of heat from the computing processes at the servers. It is important to maintain moderate temperature by employing extensive cooling in a datacenter. Otherwise, overheating and system failures will occur to disrupt the normal processing.

Cooling consumes a large amount of energy. We model the power consumed by the cooling system as $f_c^t(b(t))$, an increasing and convex function of $b(t)$. Note the function is time-dependent, as some important factors that affect the power consumption of a cooling system are time varying, such as local weather conditions.

This cooling system model is very general to capture diverse cooling facilities. As an example, the power consumption of an outdoor air cooling system can be modelled as a time-dependent cubic function of $b(t)$ [16]:

$$f_c^t(b(t)) = K_t b^3(t),$$

where $K_t > 0$ captures the ambient weather conditions, e.g., air temperature, at time t. As another example, according to [15], the power drawn from a water chiller based cooling system can be modelled as a time-dependent quadratic function of $b(t)$ as follows:

$$f_c^t(b(t)) = Q_t b^2(t) + L_t b(t) + C_t,$$

where $Q_t, L_t, C_t \geq 0$ depend on outside air and chilled water temperature at time t.

3.1.5 Power Grid Model

There are multiple possible pricing schemes for energy use. The most basic pricing scheme is flat-rate pricing, such that the energy tariff is constant, independent of time-of-use. However, flat-rate pricing does not capture the time-varying cost of energy generation. Power grid normally have various types of power generators at different generation costs (e.g., hydro, nuclear, coal-fired, gas turbine generators). Power generators with cheaper generation cost may not meet the demands at peak periods. Hence, additional power generators with higher generation cost will be also utilized during peak periods, which increases the energy generation cost considerably. Therefore, a time-of-use (ToU) pricing scheme is typically used in many countries to capture the time-varying cost of energy generation.

We assume a general model of power grid with a time-of-use (ToU) pricing scheme that dynamically varies over time. Normally, off-peak periods are cheaper than peak periods. The grid supplies energy to the datacenter in an "on-demand" fashion, with time-varying price $p(t)$ per unit energy at time t. Thus, the cost of drawing $v(t)$ units of energy from the grid at time t is $p(t) \cdot v(t)$. This model can also capture on-site renewable generation, such that $p(t)$ is the compensated price of energy consumption from renewable generation.

3.2   Problem Formulation

3.2.1   Operating Cost

Based on the above models, the total power consumption of the datacenter is the sum of the power consumed by servers, power conditioning system and cooling system. We express the total power consumption by a time-dependent function as

$$g_t(x(t), a(t)) \triangleq b(t) + f_p(b(t)) + f_c^t(b(t)),$$

where $b(t) = f_s(x(t), a(t))$ is the power consumed by servers, and $f_p(b(t))$ and $f_c^t(b(t))$ represent power consumed by power conditioning and cooling systems, respectively. We remark that $g_t(x(t), a(t))$ is increasing and convex in $x(t)$ and $a(t)$, since it is the sum of three increasing and convex functions. The results derived in this book apply to any $g_t(x(t), a(t))$ as long as it is increasing and convex in $x(t)$ and $a(t)$, for $t \in [1, T]$.

Given the workload $a(t)$, the grid price $p(t)$, and the time-dependent function g_t $(x(t), a(t))$, for $1 \leq t \leq T$, the total operating cost of the datacenter in entire horizon $[1, T]$ can be derived as follows:

$$\text{Cost}(\boldsymbol{x}, \boldsymbol{v}) \triangleq \sum_{t=1}^{T} \left(v(t)p(t) + \beta_s [x(t) - x(t-1)]^+ \right), \tag{3.1}$$

where $v(t) \geq g_t(x(t), a(t))$ is the power drawn from the grid at time t.

The cost includes the grid electricity procurement cost to power the servers, the power conditioning system, and the cooling system, as well as and the switching cost of servers. Throughout this book, we set our initial condition as $x(0) = 0$, i.e., all servers are in the off state.

3.2.2   Datacenter Capacity Provisioning Problem

Finally, we formulate the datacenter capacity provisioning problem considering operating cost minimization as the following:

$$\textbf{CP}: \quad \min \ \text{Cost}(\boldsymbol{x}, \boldsymbol{v}) \tag{3.2}$$
$$\text{s.t.} \ \ v(t) \geq g_t(x(t), a(t)), \tag{3.3}$$
$$x(t) \geq a(t), \tag{3.4}$$
$$x(0) = 0, \tag{3.5}$$
$$\text{var} \ \ x(t) \in \mathbb{N}^0, v(t) \in \mathbb{R}_0^+, t \in [1, T],$$

where

$$[\,\cdot\,]^+ = \max(0, \cdot),$$

$\mathbb{N}^0$ and $\mathbb{R}_0^+$ represent the sets of non-negative integers and real numbers, respectively.

Constraint (3.3) ensures the total power consumed by the datacenter is met by the grid electricity. Constraint (3.4) specifies that there are enough active servers to serve the workload. Constraint (3.5) is the initial condition.

The formulated problem is challenging to solve. First, it is a non-linear mixed-integer optimization problem. Further, the objective function values across different time slots are correlated via the switching costs $\beta_s[x(t) - x(t-1)]^+$ and thus cannot be solved separately. Finally, in practice we need to solve the problem in an online fashion. That is, even though the decision problems of individual time slots are coupled and should be jointly solved for optimal decisions, in practice, we can only make irrevocable decision in a sequential manner.

At each time t, we only know the inputs of the past and current time slots, but not future time slots. Furthermore, we cannot go back in time to change the decisions in previous time slots. Such "causality" constraints casted by the sequential nature of time in practice make the optimal decision making particularly challenging.

3.2.3 Discussion

We note that there are a few caveats of the problem formulation. First, the formulated problem does not consider the cost of migrating workload across servers. This is because for the mice workload model that we consider in this book, each job has a small transaction size and a short duration. Jobs arriving in a time slot will be served in the same time slot without incurring migration cost. Second, the formulated problem is similar to a common one considered in the literature, e.g., in [2]. The benefits of our particular formulation are that (i) we are able to retain the constraint that the decision variables be integers instead of real numbers and (ii) we can explicitly model the power consumption by power conditioning and cooling systems.

Note that our models are actually very general, which can be applied to other resource management scenarios. For example, deciding whether certain idling resources (e.g., manufacturing, transportation, catering resources) should be turned off and put in standby mode, considering the presence of switching costs of turning the resources on again in the future.

In the next two chapters, we first focus on developing optimal offline and online algorithms for the case with a single server. We then leverage an elegant problem structure to extend the single-server algorithms to the general case of multiple servers, without incurring optimality loss. For each case, we first design an optimal algorithm under the offline setting. By the offline setting, we mean that the workload, i.e., a, and grid electricity price, i.e., p, of all time slots in $[1, T]$, are given when making decisions in individual time slots. We then extend the algorithm to the online setting where the inputs of individual time slots are revealed sequentially and the decisions in previous time slots are irrevocable. We analyze their performance guarantees with or without (partial) future input information.

3.3   Summary

In this chapter, we defined the models of the datacenter capacity provisioning problem. In particular, we described the workload model, datacenter model, and power grid model to determine the operating cost of the datacenter. We formulated the problem in **CP**. In the next chapter, we shed light on the solution for solving the datacenter capacity provisioning in an online manner.

References

1. A. Gandhi, V. Gupta, M. Harchol-Balter, and M. Kozuch. Optimality analysis of energy-performance trade-off for server farm management. *Performance Evaluation*, 2010.
2. M. Lin, A. Wierman, L. Andrew, and E. Thereska. Dynamic right-sizing for power-proportional data centers. In *Proc. IEEE INFOCOM*, pages 1098–1106, 2011.
3. H. Qian and D. Medhi. Server operational cost optimization for cloud computing service providers over a time horizon. In *Proceedings of the 11th USENIX conference on Hot topics in management of internet, cloud, and enterprise networks and services*, pages 4–4, 2011.
4. V. Petrucci, O. Loques, and D. Mossé. Dynamic optimization of power and performance for virtualized server clusters. In *Proceedings of the 2010 ACM Symposium on Applied Computing*, pages 263–264, 2010.
5. J. Chase, D. Anderson, P. Thakar, A. Vahdat, and R. Doyle. Managing energy and server resources in hosting centers. In *Proc. ACM SOSP*, 2001.
6. M. Lin, Z. Liu, A. Wierman, and L. L. H. Andrew. Online algorithms for geographical load balancing. In *Proc. Int. Green Computing Conf.*, 2012.
7. Z. Liu, M. Lin, A. Wierman, S. Low, and L. Andrew. Greening geographical load balancing. In *Proc. ACM SIGMETRICS*, pages 233–244, 2011.
8. L. Rao, X. Liu, L. Xie, and W. Liu. Minimizing electricity cost: Optimization of distributed internet data centers in a multi-electricitymarket environment. *Proc. IEEE INFOCOM*, 2010.
9. A. Krioukov, P. Mohan, S. Alspaugh, L. Keys, D. Culler, and R. Katz. Napsac: design and implementation of a power-proportional web cluster. *ACM SIGCOMM Computer Communication Review*, 41(1):102–108, 2011.
10. E. Pinheiro, R. Bianchini, E. Carrera, and T. Heath. Load balancing and unbalancing for power and performance in cluster-based systems. In *Workshop on Compilers and Operating Systems for Low Power*, 2001.
11. R. Doyle, J. Chase, O. Asad, W. Jin, and A. Vahdat. Model-based resource provisioning in a web service utility. In *Proceedings of the 4th conference on USENIX Symposium on Internet Technologies and Systems*, 2003.
12. L. Barroso and U. Holzle. The case for energy-proportional computing. *IEEE Computer*, 40(12):33–37, 2007.
13. D. Palasamudram, R. Sitaraman, B. Urgaonkar, and R. Urgaonkar. Using batteries to reduce the power costs of internet-scale distributed networks. In *Proc. ACM Symposium on Cloud Computing*, 2012.
14. V. Mathew, R. Sitaraman, and P. Shenoy. Energy-aware load balancing in content delivery networks. In *Proc. IEEE INFOCOM*, 2012.

15. S. Pelley, D. Meisner, T. Wenisch, and J. VanGilder. Understanding and abstracting total data center power. In *Workshop on Energy-Efficient Design*, 2009.

16. Z. Liu, Y. Chen, C. Bash, A. Wierman, D. Gmach, Z. Wang, M. Marwah, and C. Hyser. Renewable and cooling aware workload management for sustainable data centers. In *Proc. ACM SIGMETRICS*, 2012.

The Case of A Single Server

We formulated the datacenter capacity provisioning problem (**CP**) in Chap. 3. Before solving the complete problem, it would be more intuitive to get the basic ideas from a simplified problem, and then extend the basic ideas to tackle the complete problem in a step-by-step manner. In this chapter, we consider a simplified case. In this case, there is only one server and toggling the server on/off (namely, setting $x(t) \in \{0, 1\}$) is the decision to make at each individual slot. We assume the workload $a(t)$ to take values in $[0, 1]$, such that it can be satisfied by one server.

In this chapter, we consider a simplified case. In this case, there is only one server and toggling it on/off is the decision to make at each individual slot. Under the setting, the single-server cost minimization problem is a simplified version of the problem **CP** in (3.2)–(3.5) as follows:

$$\textbf{SCP}: \quad \min \sum_{t=1}^{T} \left(p(t) \cdot g_t \left(x(t), a(t) \right) + \beta_s [x(t) - x(t-1)]^+ \right) \tag{4.1}$$

$$\text{s.t.} \quad x(t) \geq a(t), \forall t \in [1, T], \tag{4.2}$$

$$x(0) = 0, \tag{4.3}$$

$$\text{var} \quad x(t) \in \{0, 1\}, t \in [1, T],$$

where the workload $a(t), t \in [1, T]$, only takes values in $[0, 1]$.

Understanding the case of a single server can reveal the valuable intuition of tackling the more general multi-server problem. In particular, we can obtain a lower bound on the competitive ratio of the single-server case, which also applies to the multi-server case.

© The Author(s), under exclusive license to Springer Nature Switzerland AG 2022
M. Chen and S. C.-K. Chau, *Online Capacity Provisioning for Energy-Efficient Datacenters*, Synthesis Lectures on Learning, Networks, and Algorithms,
https://doi.org/10.1007/978-3-031-11549-3_4

4.1   Ski-Rental Approach

To illustrate the intuition of our solutions, we further focus on a special setting of **SCP** as follows:

1. The power grid price is constant, i.e.,

$$p(t) = p, \quad \forall t \in [1, T].$$

2. The power consumption function is a linear function of $x(t)$, in particular,

$$g_t(x(t), a(t)) = x(t), \quad \forall t \in [1, T].$$

3. The switching cost β_s is a positive integer. We will relax these assumptions in the subsequent sections, once we gain the intuition of our solutions.

How do we solve this simplified problem? First, since the constraint $x(t) \geq a(t)$ requires that the server needs to be active whenever there is workload, $x(t)$ has to be 1 whenever $a(t) > 0$. As such, the only non-trivial decision in solving the problem **SCP** is to determine when to set $x(t) = 0$ after $a(t) = 0$. More precisely, we need to decide when to turn off the server if there is no workload.

The decision-making of turning the server off is a non-trivial problem. On one hand, keeping the server idling will incur idling cost linearly proportional to the length of the idling period. On the other hand, turning off the server too immediately (as soon as no workload arrives) will incur zero idling cost, but a potentially substantial turning-on cost if there is further workload in the next slots. It is evident that this is an online "ski-rental"-like decision problem of whether to "buy", i.e., turning off the serve now and turning it on upon next workload arrival, or to "rent", i.e., keeping the server idle. However, there is a notable difference with the classical ski-rental problem—our problem of a single server is actually a "repeated" ski-rental problem, in which each ski-rental instance corresponds to whether to turn off the active server upon the time epoch of no workload arrival. This observation inspires our approach to solve our problem as follows.

4.1.1   Offline Solution

We first examine the offline solution without uncertainty of future workload. In the offline setting, $a(t) \in [0, 1]$ and $p > 0, t \in [1, T]$, are given in advance. Hence, one can aggregate the offline optimal solutions for individual ski-rental instances to obtain an optimal offline solution for the single-server cost-minimization problem. At each time t, we can find the offline optimal solutions by the following cases:

- **Case 1**: If the server is off and $a(t) > 0$, i.e., there is a workload, then turn on the server. Otherwise, it will violate the constraint $x(t) \geq a(t)$.
- **Case 2**: If the server is on and $a(\tau) = 0$ for all $\tau \in \left[t, t + \frac{\beta_s}{p}\right]$, i.e., no workload arrival in the next $\frac{\beta_s}{p}$ amount of time, then turn off the server. Note that if the server remains on with no workload for a duration of $\frac{\beta_s}{p}$, there incurs a total cost $\frac{\beta_s}{p} p = \beta_s$. Hence, it is more sensible to turn off the server from time t.
- **Case 3**: In all other cases, keep the server 'on/off' state unchanged.

It is evident that this will make an offline optimal solution, similar to that of the ski-rental problem.

4.1.2 Online Solution

We then examine the online solution without knowing the future workload. In the online setting, $a(t), t \in [1, T]$, is revealed in a sequential manner and the past decisions are irrevocable. We can extend the optimal deterministic and randomized online algorithms for individual ski-rental instances to obtain optimal online algorithms for our problem. Specifically, we can find the online optimal solutions by the following cases:

- **Case 1**: Turn on the server (or keep it on/active if it is already so) whenever there is workload.
- **Case 2**: Upon no workload arrival and the server is in an idle (active) state, wait for Δ slots before turning off the server, if no further workload arrives.

When applying the deterministic online algorithm of the ski-rental problem, the value of Δ is set to be $\frac{\beta_s}{p}$, which is regarded as a break-even interval comparing with $\frac{\beta_s}{p}$. When applying the randomized online algorithm of the ski-rental problem, the value of Δ is generated according to a probability distribution as shown in Chap. 2, Sect. 2.3.4 in the respective randomized algorithm. As you would expect, it is straightforward to show that 2 and $\frac{e}{1-e}$ are the best possible competitive ratios for deterministic and randomized online algorithms for our single-server cost-minimization problem, respectively.

Note that one might suspect that, since the latter is a repeated version of the former one, there might exist a larger design space of online algorithms than that of individual ski-rental problem and a better competitive ratio might be possible. This conjecture makes sense but one can show that the competitive ratios cannot be improved, by induction. For the first ski-rental instance, it is clear that there exists a worst-case input where 2 and $\frac{e}{1-e}$ are the best achievable ratios. Then assuming the claim holds for the first k instances, we argue that for $k + 1$ instances, we can expand the previous worst-case input by one more worst-case segment for the $(k + 1)$-th instance. Consequently, 2 and $\frac{e}{1-e}$ are the best achievable ratios for $k + 1$ instances as well and thus the claim holds for the repeated ski-rental problem.

4.1.3   Online Solution with Prediction

Let us also understand how prediction can benefit our single-server cost-minimization problem. It is evident that (perfect) prediction can help to improve performance of online algorithms and hence the competitive ratios. Specifically, if the server operator can predict the workload w slots ahead, i.e., $a(\tau)$ for $\tau \in [t, t+w]$ are known at time t, then the operator upon solving each ski-rental instance can turn off the idling server w slots earlier than the case without prediction. In this way, the behavior of an online algorithm is more aligned with that of the offline algorithm, resulting in a better competitive ratio.

4.2   Optimal Offline Algorithm

After gaining the intuition from the ski-rental approach, we consider a more general case with time-varying grid price $p(t)$ and arbitrary $g_t(x(t), a(t))$ that is an increasing and convex function. Equipped with the insights from the previous section, we develop an optimal offline algorithm for problem **SCP** in Algorithm 4.1.

Algorithm 4.1 Optimal offline algorithm **OffSCP**

Input : $a(t) \in [0, 1]$, $p(t)$, and $g_t(\cdot, \cdot)$, for all $t \in [0, T]$
Output : $x(t) \in \{0, 1\}$, $t \in [1, T]$
Initialization : $x(0) \leftarrow 0$

1: **for** $t = 1$ to T **do**
2:    **if** $a(t) > 0$ **then**
3:       $x(t) \leftarrow 1$
4:    **else if** $a(t) = 0$ and $a(t - 1) = 1$ **then**
5:       Find $t' > t$ such that $a(t') > 0$ for the first time after t
6:       **if** $\sum_{\tau=t}^{t'-1} p(\tau)\big(g_\tau(1, 0) - g_\tau(0, 0)\big) < \beta_s$ **then**
7:          $x(t) \leftarrow 1$ // *keep server on*
8:       **else**
9:          $x(t) \leftarrow 0$ // *switch server off*
10:      **end if**
11:   **else**
12:      $x(t) \leftarrow x(t - 1)$
13:   **end if**
14: **end for**
15: Return $x(t)$ for all $t \in [1, T]$

The intuition obtained from the ski-rental approach still applies, but with adjustment in setting the "break-even" criterion. In particular, upon entering idling state, instead of comparing the server idling time to the break-even interval, we now compare the aggregate idling cost under time-varying grid prices to the switching cost (i.e., $\sum_{\tau=t}^{t'-1} p(\tau)\big(g_\tau(1,0) - g_\tau(0,0)\big) \leq \beta_s$), and turn off the server when the former equals or just exceeds the latter. The optimality of **OffSCP** (Algorithm 4.1) is shown in Proposition 4.2. We first provide the definition of monotonically increasing function, which will be used to describe the nature property of the energy consumption function concerned in Proposition 4.2.

Definition 4.1 A function $\xi(x, y) : \mathbb{R}^2 \to \mathbb{R}$ is monotonically increasing if for $x_2 \geq x_1$ and $y_2 \geq y_1$, we have $\xi(x_2, y_2) \geq \xi(x_1, y_1)$.

Proposition 4.2 *Consider monotonically increasing functions $g_t(x(t), a(t))$, $t \in [1, T]$. In the offline setting where $a(t) \in [0, 1]$, $p(t)$, and $g_t(x(t), a(t))$, for all $t \in [0, T]$, are given in advance, **OffSCP** (Algorithm 4.1) produces an optimal solution to the problem **SCP**.*

Proof We first define four types of intervals to facilitate the discussion:

- **Idling Interval**: $I_1 \triangleq [t_1, t_2]$ is called an idling interval, if $a(t_1 - 1) > 0$, $a(t_2 + 1) > 0$, and $a(\tau) = 0$ for all $\tau \in [t_1, t_2]$.
- **Working Interval**: $I_2 \triangleq [t_1, t_2]$ is called a working interval, if $a(t_1 - 1) = 0$, $a(t_2 + 1) = 0$, and $a(\tau) > 0$ for all $\tau \in [t_1, t_2]$.
- **Starting Interval**: $I_s \triangleq [0, t_2]$ is called a starting interval, if $a(t_2 + 1) > 0$ and $a(\tau) = 0$ for all $\tau \in [0, t_2]$.
- **Ending Interval**: $I_e \triangleq [t_1, T + 1]$ is called a starting interval, if $a(t_1 - 1) > 0$ and $a(\tau) = 0$ for all $\tau \in [t_1, T + 1]$.

It is evident that it is optimal to set $x = 0$ during I_s and I_e and set $x = 1$ during each I_2. During an idling interval denoted as I_1, an offline optimal solution must set $x(\tau)$ to be always 1 or 0, $\forall \tau \in I_1$; otherwise, toggling a server on/off in the middle of the interval will incur unnecessary switching cost, which is sub-optimal.

Note that cost corresponding to $x(\tau) = 1$, $\forall \tau \in I_1$, is $\sum_{\tau \in I_1} p(\tau) g_\tau(1, 0)$. The cost corresponding to $x(\tau) = 0$, $\forall \tau \in I_1$, is β_s, since we must pay a switching cost β_s at the end of the interval. It is thus optimal to compare the two costs and choose the most economic solution for this interval I_1.

The above argument shows that the solution generated by **OffSCP** is thus an optimal offline solution. □

The complexity of Algorithm 4.1 scales as $O(T)$, as one only needs to perform simple summations and comparisons in each slot.

4.3   Deterministic Online Algorithm

We next extend the online algorithms for the ski-rental problem to a more general case with time-varying grid price $p(t)$ and arbitrary $g_t(x(t), a(t))$. A deterministic online algorithm for problem **SCP** is described in **OnSCP** (Algorithm 4.2). In this algorithm, we record the cumulative "idling cost" C since the last time when the server has become idling with no workload. Similar to the break-even algorithm, the server will be switched off, when the cumulative idling cost C exceeds the switching cost β_s.

Algorithm 4.2 Deterministic online algorithm **OnSCP**

Input : $a(t) \in [0, 1]$, $p(t)$, and $g_t(\cdot, \cdot)$, at current time t
Output : $x(t) \in \{0, 1\}$, at current time t
Initialization : $C \leftarrow 0$, $x(0) \leftarrow 0$

1: **if** $a(t) > 0$ **then**
2:    $x(t) \leftarrow 1$, $C \leftarrow 0$ // *reset cumulative idling cost*
3: **else if** $C < \beta_s$ and $a(t) = 0$ **then**
4:    $x(t) \leftarrow x(t-1)$, $C \leftarrow C + p(t)\big(g_t(1, 0) - g_t(0, 0)\big)$
5: **else**
6:    $x(t) \leftarrow 0$, $C \leftarrow 0$ // *switch server off*
7: **end if**
8: Return $x(t)$

Proposition 4.3 *Consider monotonically increasing functions* $g_t(x(t), a(t))$, $t \in [1, T]$. *Under the online setting where at time t, $a(\tau) \in [0, 1]$, $p(\tau)$, and $g_\tau(x(\tau), a(\tau))$, for all $\tau \in [0, t]$, are available, **OnSCP** (Algorithm 4.2) achieves the best possible competitive ratio 2 among all deterministic online algorithms for **SCP**.*

It suffices to consider only idling intervals I_1. It is evident that the cumulative idling cost $\sum_{t \in I_1} p(t)\big(g_t(1, 0) - g_t(0, 0)\big)$ is equivalent to the renting cost in the ski-rental problem. Hence, we can apply the similar argument of the deterministic online algorithm in the ski-rental problem **OnSki** (Algorithm 2.1) for **OnSCP** (Algorithm 4.2) to prove its competitive ratio. We skip the proof as it is similar to that of the deterministic online algorithms for the ski-rental problem.

4.4 Deterministic Prediction-Aware Online Algorithm

Following the intuitions in Section 4.1.3, we develop a deterministic prediction-aware online algorithm for problem **SCP**, taking prediction into account.

Under the online setting, the inputs $a(t)$, $p(t)$, and $g_t(x(t), a(t)), t \in [1, T]$, are revealed in a sequentially manner and past decisions are irrevocable. Moreover, the prediction information is available in the sense that at time t, we also know $\{a(\tau), p(\tau), g_\tau(x(t), a(t)), \forall \tau \in [t, t + \omega]\}$, where $\omega \geq 0$ is the duration of prediction window. The case of $\omega = 0$ corresponds to no prediction. A prediction-aware online algorithm is described in **PrOnSCP** (Algorithm 4.2). **PrOnSCP** is a simple extension of **OnSCP** by switching the server off earlier by looking ahead to see if the cumulative idling cost exceeds the switching cost in the prediction window.

Algorithm 4.3 Deterministic prediction-aware online algorithm **PrOnSCP**

Input : $a(t) \in [0, 1]$, $p(t)$, $g_t(\cdot, \cdot)$, at current time t, and prediction window ω
Output : $x(t) \in \{0, 1\}$, at current time t
Initialization : $C \leftarrow 0, x(0) \leftarrow 0$

1: // check if cumulative idling cost exceeds β_s in prediction window
2: $\tau' \leftarrow \min \left\{ t' \in [t, t + \omega] \mid C + \sum_{\tau=t}^{t'} p(\tau)\big(g_\tau(1, 0) - g_\tau(0, 0)\big) > \beta_s \right\}$
3: **if** $a(t) > 0$ **then**
4:     $x(t) \leftarrow 1$ and $C \leftarrow 0$
5: **else if** $\tau' = \text{Null}$ or $\exists \tau \in [t, \tau'], a(\tau) > 0$ **then**
6:     $x(t) \leftarrow x(t - 1), C \leftarrow C + p(t)\big(g_t(1, 0) - g_t(0, 0)\big)$
7: **else**
8:     $x(t) \leftarrow 0, C \leftarrow 0$
9: **end if**
10: Return $x(t)$

The following theorem shows that **PrOnSCP** achieves the best possible competitive ratio among all deterministic algorithms, potentially with prediction taken into design consideration.

Theorem 4.4 *Consider arbitrary $g_t(x(t), a(t)), t \in [1, T]$, that are increasing and convex. Under the online setting where at time t, $a(\tau) \in [0, 1]$, $p(\tau)$, and $g_\tau(x(\tau), a(\tau))$, for all $\tau \in [0, t + \omega]$, are available, **PrOnSCP** (Algorithm 4.2) achieves the best possible competitive ratio $2 - \alpha_s$ among all deterministic online algorithms for **SCP**. Here*

$$\alpha_s \triangleq \min\left(1, \omega \cdot P_{min} d_{min} \frac{1}{\beta_s}\right) \in [0, 1],$$

is a "normalized" prediction window duration and

$$d_{\min} \triangleq \min_{t \in [1,T]} \left\{ g_t(1,0) - g_t(0,0) \right\}. \tag{4.4}$$

We defer the full proof in the Appendix.

Remarks: We elaborate the following observations from the results in Theorem 4.4. First of all, when $\omega = 0$ and there is no prediction available, **PrOnSCP** reduces to the classical break-even algorithm and the corresponding competitive ratio is 2. Meanwhile, when $\omega > 0$, **PrOnSCP** is able to exploit the prediction to achieve a competitive ratio of $2 - \alpha_s$, strictly better than the one without prediction. In particular, the competitive ratio decreases *linearly* in the prediction window duration ω until α_s increases to one, as long as $d_{\min}$ is positive.

- If $g_t(x,0)$ is strictly increasing in x, we must have $d_{\min} > 0$. In this case, the competitive ratio decreases as the prediction window duration ω increases. Further, when the prediction window duration ω reaches a break-even interval $\Delta_s \triangleq \frac{\beta_s}{d_{\min} P_{\min}}$, Algorithm 4.2 achieves a competitive ratio of 1, and further increasing the prediction window will not decrease the cost. Intuitively, this is because when knowing the inputs in $[t, t + \Delta_s]$ is sufficient for Algorithm 4.2 to behave exactly the same as the offline optimal algorithm and achieve the minimum possible cost.
- If $g_t(x,0)$ is not strictly increasing in x (*e.g.*, when there is enough renewable energy to satisfy the entire datacenter energy demand and thus $g_t(x,0)$ always takes value zero), then $d_{\min}$ is 0. In this case, the competitive ratio of **PrOnSCP** does not decrease as ω increases (but remains 2). However, the competitive ratio 2 is for the worst case inputs. In practice, **PrOnSCP** can still benefit from prediction. As long as power consumption $g_t(x,a)$ is not always non-negative. Then, for the server, with look-ahead, **PrOnSCP** can tell whether the cumulative idling cost will reach β_s earlier than without prediction. In this way, it benefits from future information.

Finally, based on the empirical evaluation in Chap. 6, we observe that $g_t(x,a)$ is strictly increasing in x and is always non-negative. Thus, $d_{\min} > 0$ is positive in practical scenarios.

4.5   Randomized Online Algorithm

In addition to the deterministic online algorithm, we also derive a randomized online algorithm for problem **SCP**. Our randomized online algorithm is described in **ROnSCP** (Algorithm 4.4). The idea of **ROnSCP** is to follow that of **ROnSki** (Algorithm 2.2) in the ski-rental problem.

In this algorithm, rather than setting the decision criterion at the switching cost β_s, it randomly generates a threshold of cumulative idling cost Λ, which is less than β_s. More specifically, **ROnSCP** for problem **SCP** performs as follows: it accumulates the idling cost and when it is less than threshold Λ, which is a random variable to be introduced later, it keeps the server idling; otherwise, it will see whether the workload will arrives, *i.e.*, $a_i > 0$, before the "idling cost" reaches β_s. If so, it keeps the server idling; else it turns off the server.

Specifically, we define the random variable Λ by the probability distribution function $f_\Lambda(\lambda)$ as follows:

$$f_\Lambda(\lambda) \triangleq \begin{cases} \frac{e^{\lambda/[\beta_s]}}{\beta_s(e-1)}, & \text{if } 0 \le \lambda \le \beta_s, \\ 0, & \text{otherwise,} \end{cases} \tag{4.5}$$

Algorithm 4.4 Randomized online algorithm **ROnSCP**

Input : $a(t) \in [0, 1]$, $p(t)$, and $g_t(\cdot, \cdot)$, at current time t
Output : $x(t) \in \{0, 1\}$, at current time t
Initialization : $C \leftarrow 0, x(0) \leftarrow 0$

1: // *randomly generate a threshold of cumulative idling cost*
2: Λ is randomly generated according to $f_\Lambda(\lambda)$
3: **if** $a(t) > 0$ **then**
4:    $x(t) \leftarrow 1, C \leftarrow 0$
5: **else if** $C < \Lambda$ and $a(t) = 0$ **then**
6:    $x(t) \leftarrow x(t-1), C \leftarrow C + p(t)\big(g_t(1, 0) - g_t(0, 0)\big)$
7: **else**
8:    $x(t) \leftarrow 0, C \leftarrow 0$
9: **end if**
10: Return $x(t)$

Theorem 4.5 *ROnSCP (Algorithm 4.4) for problem **SCP** has a competitive ratio of $\frac{e}{e-1}$. Further, no randomized online algorithm can achieve a smaller expected competitive ratio.*

It suffices to consider only idling intervals I_1. We can apply the similar argument of the randomized online algorithm in the ski-rental problem **ROnSki** (Algorithm 2.2) for **ROnSCP** (Algorithm 4.2) to prove its expected competitive ratio. We skip the proof as it is similar to that of the randomized online algorithms for the ski-rental problem.

4.6   Randomized Prediction-Aware Online Algorithm

ROnSCP can be extended to take prediction into consideration. The prediction information is available in the sense that at time t, we also know $\{a(\tau), p(\tau), g_\tau(x(t), a(t)), \forall \tau \in [t, t+\omega]\}$. The case of $\omega = 0$ corresponds to no prediction. A randomized prediction-aware online algorithm is described in **RPrOnSCP** (Algorithm 4.5). Note that the threshold Λ follows a different probability distribution defined as follows:

$$f_\Lambda^{\alpha_s}(\lambda) \triangleq \begin{cases} \frac{e^{\lambda/[\beta_s(1-\alpha_s)]}}{\beta_s(1-\alpha_s)(e-1+\alpha_s)} + \frac{\alpha_s}{(e-1+\alpha_s)}\delta(\lambda), & \text{if } 0 \leq \lambda \leq (1-\alpha_s)\beta_s, \\ 0, & \text{otherwise,} \end{cases} \tag{4.6}$$

where $\alpha_s = \min\{1, \omega d_{\min} P_{\min}\frac{1}{\beta_s}\}$ and $\delta(\lambda)$ is Dirac Delta function.

Fig. 4.1   PDF $f_\Lambda^{\alpha_s}(\lambda)$. Here $\alpha_s = 0.4$

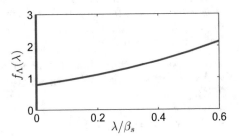

We remark that $f_\Lambda^{\alpha_s}(\lambda) = f_\Lambda(\lambda)$ when $\alpha_s = 0$ (namely, $\omega = 0$). That is, this reduces to the case without prediction, and **RPrOnSCP** becomes **ROnSCP** (Fig. 4.1).

Theorem 4.6 *RPrOnSCP (Algorithm 4.5) for problem SCP has a competitive ratio of $\frac{e}{e-1+\alpha_s}$, where $\alpha_s = \min\{1, \omega d_{\min} P_{\min}\frac{1}{\beta_s}\}$. Further, no randomized online algorithm with a prediction window ω can achieve a smaller competitive ratio.*

We defer the full proof in the Appendix.

Remarks: A consequence of Theorem 4.6 is that when the prediction window duration ω reaches a break-even interval $\Delta_s \triangleq \frac{\beta_s}{d_{\min} P_{\min}}$, **RPrOnSCP** has a expected competitive ratio of 1, similar to **PrOnSCP**.

In addition, Theorems 4.4 and 4.6 imply that the worst-case performances of our prediction-aware online algorithms are determined by how much we can see into the future. The online algorithms perform better in the worst-case with a larger prediction window. In practice, we have similar observations: the empirical performances (to be shown in Chap. 6)

Algorithm 4.5 Randomized prediction-aware online algorithm **RPrOnSCP**

Input : $a(t) \in [0, 1]$, $p(t)$, and $g_t(\cdot, \cdot)$, at current time t
Output : $x(t) \in \{0, 1\}$, at current time t
Initialization : $C \leftarrow 0$, $x(0) \leftarrow 0$

1: Λ is randomly generated according to $f_\Lambda^{\alpha_s}(\lambda)$
2: // *check if cumulative idling cost exceeds* Λ *in prediction window*
3: $\tau' \leftarrow \min \left\{ t' \in [t, t + \omega] \mid C + \sum_{\tau=t}^{t'} p(\tau)\big(g_\tau(1, 0) - g_\tau(0, 0)\big) > \Lambda \right\}$
4: **if** $a(t) > 0$ **then**
5:   $x(t) \leftarrow 1$, $C \leftarrow 0$
6: **else if** $\tau' = \texttt{Null}$ or $\exists \tau \in [t, \tau'], a(\tau) > 0$ **then**
7:   $x(t) \leftarrow x(t - 1)$, $C \leftarrow C + p(t)\big(g_t(1, 0) - g_t(0, 0)\big)$
8: **else**
9:   $x(t) \leftarrow 0$, $C \leftarrow 0$
10: **end if**
11: Return $x(t)$

---

of our algorithms quickly improve to the offline optimal as prediction window duration ω increases. This indicates that the value of short-term prediction of inputs can dramatically reduce the operating cost of the datacenter.

4.7   Summary

In this chapter, we considered a simplified case with only one server in problem **SCP**. We presented deterministic online algorithms with and without prediction. We also presented randomized online algorithms with and without prediction. We defer the proofs of the competitive ratios of these online algorithms to the subsequent chapters. We will apply our algorithms to the multi-server case in the next chapter.

The General Case of Multiple Servers

After formulated the datacenter capacity provisioning problem (**CP**) in Chaps. 3, 4 studied
a simplified case with only one server (**SCP**). In this chapter, we generalize the ideas of
the single-server case to the multi-server case. To do so, we apply a slicing technique to
divide the multi-server problem into multiple single-server subproblems. We then solve each
subproblem separately. We combine the sub-problem solutions to obtain a solution for the
overall problem. The key is to correctly decompose the demand and define the subproblems
so that (i) the combined solution is optimal to the overall problem under the offline setting
and (ii) the demand decomposition can be done in an online manner and thus the subproblems
can also be solved under the online setting with strong performance guarantee.

5.1   Workload Decomposition and Cost Minimization Subproblems

Under the multi-server setting, the total workload may exceed the processing capability of
a single server. Thus it is necessary to construct and distribute sub-workloads to individual
servers to serve. Specifically, we first "slice" the demand a into sub-demands, define a
sub-problem associated with each sub-demand for each server, and then solve capacity
provisioning *separately* for each server. More specifically, we slice the demand as follows.
Let $M = \max_{1 \leq t \leq T} \lceil a(t) \rceil$. For $1 \leq i \leq M$ and $1 \leq t \leq T$, we define

$$a_i(t) \triangleq \min \left\{ 1, \max \left\{ 0, a(t) - (i-1) \right\} \right\}. \tag{5.1}$$

Namely, $\{a_i(t)\}$ is a decomposition of workload $a(t)$, i.e., $a(t) = \sum_{i=1}^{M} a_i(t)$. An example of the decomposed workload is shown in Fig. 5.1. We note that this particular way of
constructing the sub-workload have the following advantages. First, the decomposition of
$a(t)$ can be performed in an online fashion as it does not utilize the knowledge of future

© The Author(s), under exclusive license to Springer Nature Switzerland AG 2022
M. Chen and S. C.-K. Chau, *Online Capacity Provisioning for Energy-Efficient
Datacenters*, Synthesis Lectures on Learning, Networks, and Algorithms,
https://doi.org/10.1007/978-3-031-11549-3_5

Fig. 5.1 An example of how
workload a is decomposed into
4 sub-demands

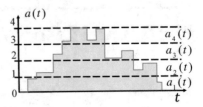

workload. Second, intuitively, the sub-workloads obtained according to (5.1) are most unbalanced, which allow some servers to stay as busy as possible thus others can be turned off
during the time interval without sub-workload to reduce the overall cost. It turns out that
such unbalanced sub-workload and the busy-servers-get-busier approach help to minimize
the number of toggling on/off times and thus the total cost.

Given the decomposed sub-workloads $\{a_i(t)\}$, we allocate each sub-workload to a server
and solve the following sub-problems $\mathbf{CP}_i$, $1 \leq i \leq M$, defined as follows:

$$\mathbf{CP}_i : \min \sum_{t=1}^{T} \left\{ p(t) \cdot d_i(t) \cdot x_i(t) + \beta_s [x_i(t) - x_i(t-1)]^+ \right\}$$
$$\text{s.t. } x_i(t) \geq a_i(t),$$
$$x_i(0) = 0,$$
$$\text{var } x_i(t) \in \{0, 1\}, \ t \in [1, T].$$

Here, $x_i(t)$ indicates whether the i-th server is on at time t, and

$$d_i(t) \triangleq g_t(i, a(t)) - g_t(i-1, a(t))$$

represents the power consumption increment due to using the i-th server at t. Problem
$\mathbf{CP}_i$ represents the capacity provisioning problem with inputs workload a_i, grid price p and
$d_i(t)$. We note that, being a single server problem with workload a_i in $[0, 1]$, the subproblems
$\mathbf{CP}_i$, $1 \leq i \leq M$, are easier to solve than the problem $\mathbf{CP}$. Indeed, each $\mathbf{CP}_i$ is simply the
single server problem $\mathbf{SCP}$ studied in the previous chapter. It can be solved optimally by
Algorithm 4.3 under the offline setting.

Next, we discuss how one can combine the solutions to $\mathbf{CP}_i$ to obtain an optimal solution
to $\mathbf{CP}$. At any time t,

$$x_i(t) \geq a_i(t), \ \ 1 \leq i \leq M;$$

thus we have

$$\sum_{i=1}^{M} x_i(t) \geq \sum_{i=1}^{M} a_i(t) = a(t).$$

and the combined solution is feasible for the problem $\mathbf{CP}$.

5.1.1 Workload Decomposition Incurs No Optimality Loss Under the Offline Setting

It should be clear that, under the offline setting, the solution obtained by the "divide-and-conquer" approach is sub-optimal in general. It explores only a subset of the design space of the original problem **CP**, by specifying a special structure of serving the overall workload and as such the obtained solution. Such optimality loss under the offline setting can be termed as "price of decomposition". One of our contributions is to show that, interestingly, as the following theorem states, the individual optimal offline solutions for problems **CP**$_i$ can be put together to obtain an optimal offline solution to the original problem **CP**. Hence, the incurred price-of-decomposition is zero.

Theorem 5.1 *Consider monotonically increasing functions* $g_t(x(t), a(t))$, $t \in [1, T]$. *Let* x_i^* *be an optimal offline solution for problem* **CP**$_i$ *with workload* a_i, $1 \le i \le M$. *Then* $\sum_{i=1}^{M} x_i^*$ *is an optimal offline solution for* **CP** *with workload* a.

The theorem essentially says that the most unbalanced sub-workloads obtained by (5.1) and individual servers solving their own sub-problems independently are cost-minimizing. Intuitively, the unbalanced sub-workloads allow some servers to stay as busy as possible thus the remaining servers can be turned off during the idling intervals to minimize the overall cost.

5.1.2 Online Algorithm Design Following the Divide-and-Conquer Approach

There is a natural idea to design an online algorithm following the "divide-and-conquer" approach, as suggested by Theorem 5.1 and made clear in the corollary below.

Corollary 5.1.1 *If an online algorithm can achieve a competitive ratio of* γ *for the sub-problem* **CP**$_i$, *then applying the algorithm to all sub-problems* **CP**$_i$ *with workload* a_i, $1 \le i \le M$, *and combining the online solutions gives an online solution to problem* **CP** *with workload* a *with a competitive ratio* γ.

The corollary is a direct consequence of Theorem 5.1 and that the sub-workloads a_i, $1 \le i \le M$ can be obtained in an online fashion. In the next, we will present the optimal offline and online algorithms for the multi-server case.

5.2   Optimal Offline and Online Algorithms

Given any feasible workload a where $0 \leq a(t) \leq M, 1 \leq t \leq T$, the following offline algorithm generates a solution to the problem **CP**.

Algorithm 5.1 Optimal offline algorithm under the multi-server setting **OffCP**

Input : $a(t) \in [0, M]$, $p(t)$, and $g_t(\cdot, \cdot)$, for all $t \in [0, T]$
Output : $x(t) \in \{0, 1\}, t \in [1, T]$
Initialization : $x(0) \leftarrow 0$

1: **for** $t = 1$ to T **do**
2:    Obtain $a_i(t)$, $1 \leq i \leq M$, by (5.1)
3: **end for**
4: **for** $i = 1$ to M **do**
5:    Obtain x_i by calling **OffSCP**$(a_i, p, g(\cdot, \cdot))$
6: **end for**
7: Return $x(t) = \sum_{i=1}^{M} x_i(t)$ for all $t \in [1, T]$

The algorithm exactly obtains the most unbalanced sub-workloads, assigns each of them to a server, solves a single server problem with the sliced sub-workloads optimally, and combines the obtained solutions to generate one for the overall problem. The following proposition confirms its optimality.

Proposition 5.2.1 *Consider arbitrary $g_t(x(t), a(t))$, $t \in [1, T]$, that are increasing and convex. The solution obtained by Algorithm **OffCP** is optimal for the problem **CP**.*

The above proposition is a direct consequence of Theorem 5.1 and that Algorithm **OffSCP** is optimal for the single server problem **SCP**.

For online algorithms, we also apply then divide-and-conquer approach by using (i) a central demand dispatching module that slices and dispatches workload to individual servers according to (5.1), and (ii) an online scheduling module sitting on each server independently solving their own single server problem using the deterministic and randomized online algorithms discussed in the previous chapter. We summarize the resulting algorithm, with prediction information taken into account, in the following.

The online algorithms, named **PrOnCP**(ω) and **RPrOnCP**(ω), are simple to implement without the need to coordinate the control among multiple servers. Based on the discussions in the previous section, it is straightforward to establish their competitive ratios.

Proposition 5.2.2 *Consider arbitrary $g_t(x(t), a(t))$, $t \in [1, T]$, that are increasing and convex. Under the online setting where at time t, $a(\tau) \in [0, M]$, $p(\tau)$, and $g_\tau(x(\tau), a(\tau))$,*

Algorithm 5.2 Deterministic prediction-aware online algorithm **PrOnCP**(ω) and **RPrOnCP**(ω)

Input : $a(t) \in [0, M]$, $p(t)$, $g_t(\cdot, \cdot)$, at current time t, and prediction window ω
Output : $x(t) \in \{0, 1\}$, at current time t
Initialization : $C \leftarrow 0$, $x(0) \leftarrow 0$

1: **for** $\tau = t$ to $t + \omega$ **do**
2:    Obtain $a_i(\tau)$, $1 \le i \le M$, by (5.1)
3: **end for**
4: **for** $i = 1$ to M **do**
5:    For **PrOnCP**(ω): Obtain $x_i(t)$ by calling **PrOnSCP**($a_i(\tau)$, $p(\tau)$, $g_\tau(\cdot, \cdot)$, $\tau \in [t, t + \omega]$)
6:    For **RPrOnCP**(ω): Obtain $x_i(t)$ by calling **RPrOnSCP**($a_i(\tau)$, $p(\tau)$, $g_\tau(\cdot, \cdot)$, $\tau \in [t, t + \omega]$)
7: **end for**
8: Return $x(t) = \sum_{i=1}^{M} x_i(t)$

---

*for all $\tau \in [0, t + \omega]$, are available, Algorithm **PrOnCP**(ω) achieves a competitive ratio of $2 - \alpha_s$ for the problem **CP**. Algorithm **RPrOnCP**(ω) achieves a competitive ratio of $\frac{e}{e-1+\alpha_s}$. Here,*

$$\alpha_s \triangleq \min\left(1, \omega \cdot P_{\min} d_{\min} \frac{1}{\beta_s}\right) \in [0, 1],$$

is a "normalized" prediction window duration and

$$d_{\min} \triangleq \min_{t \in [1,T]} \min_{1 \le i \le M} \left\{g_t(i, 0) - g_t(i - 1, 0)\right\}. \tag{5.2}$$

*Furthermore, the achieved competitive ratios are optimal for all deterministic and randomized online algorithms for the problem **CP**.*

The above proposition is a direct consequence of Theorem 5.1 and that Algorithm **PrOn-SCP** and **RPrOnSCP** are optimal online algorithms for the single server problem **SCP**.

5.3   Discussions

We summarize the comparison of representative existing works, in particular LCP and **PrOnCP/RPrOnCP**, on the topic in Table 5.1). As seen from the table, from the modeling aspect, **PrOnCP/RPrOnCP** explicitly take into account power consumption of both cooling and power conditioning systems, in addition to servers, and power supply from renewable energy sources. From the formulation aspect, **PrOnCP/RPrOnCP** are solving a different optimization problem, i.e., an integer program with convex and increasing objective function. From the theoretical result aspect, **PrOnCP** and **RPrOnCP** achieve competitive ratios of

Table 5.1 Summary of the differences between algorithms that we developed in this monograph and previous ones. Note: Here Pow. Cond. stands for Power Conditioning. C.R. stands for Competitive Ratio. Deter. stands for Deterministic. Rand. stands for randomized. Alg. stands for Algorithm. α_s is the normalized look-ahead window size

	Pow. cond.,	Optimization	Problem	C.R.	
	Cooling Renewable	Objective Function	Variable Type	Deter. Alg.	Rand. Alg.
LCP [1]	No	Convex	Continuous	3	$\times$
OnCP/ROnSCP (this monograph)	Yes	Convex and increasing	Integer	$2 - \alpha_s$	$\frac{e}{e-1+\alpha_s}$

$2 - \alpha_s$ and $e/(e - 1 + \alpha_s)$, respectively, where α_s is the normalized look-ahead window size. Both ratios quickly decrease to 1 as the look-ahead window w increase.

Reference

1. M. Lin, A. Wierman, L. Andrew, and E. Thereska. Dynamic right-sizing for power-proportional data centers. In *Proc. IEEE INFOCOM*, pages 1098–1106, 2011.

We implemented the proposed offline and online algorithms and carry out simulations using real-world traces to evaluate their performance. Our aims are threefold. First, to evaluate the performance of the algorithms in a typical setting. Second, to study the impacts of workload prediction error and workload characteristics on the algorithms' performance. Third, to compare our algorithms to two recently proposed solutions **LCP**(ω) in [1] and **DELAYEDOFF** in [2].

6.1 Settings

Workload trace: The real-world traces we use in experiments are a set of I/O traces taken from 6 RAID volumes at MSR Cambridge [3]. The traced period was one week from February 22 to 29, 2007. We estimate the average number of jobs over disjoint 10 min intervals. The data trace has a peak-to-mean ratio (PMR) of 4.63. The jobs are "request-response" type and thus the workload is better described by a discrete-time fluid model, with the slot length being 10 min and the load in each slot being the average number of jobs.

In the experiments, we run algorithm **LCP**(ω) [1] by directly using the above discrete-time trace, since **LCP**(w) was originally designed to work under a discrete-time setting. Meanwhile, **PrOnSCP**, **RPrOnSCP**, and **DELAYEDOFF** [2] were primally designed to work under a continuous-time setting. To evaluate their performance by using the above discrete-time trace, we run these algorithms by feeding jobs continuously to the algorithms, where the job-arrivals in a slot are assumed to uniformly spread out the slot. By this setting, we would like to demonstrate that algorithms **PrOnSCP/RPrOnSCP/DELAYEDOFF** do not require to know the number of job-arrivals a priori to operate. We use last-empty-server-first job-dispatching strategy for **RPrOnSCP** (Fig. 6.1).

Cost benchmark: Current datacenters usually do not use dynamic provisioning. The cost incurred by static provisioning is usually considered as benchmark to evaluate new

© The Author(s), under exclusive license to Springer Nature Switzerland AG 2022   51
M. Chen and S. C.-K. Chau, *Online Capacity Provisioning for Energy-Efficient Datacenters*, Synthesis Lectures on Learning, Networks, and Algorithms,
https://doi.org/10.1007/978-3-031-11549-3_6

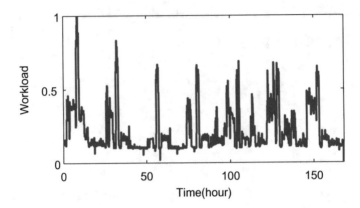

Fig. 6.1 Trace for one week in MSR

algorithms [1, 4]. Static provisioning runs a constant number of servers to serve the workload. In order to satisfy the time-varying demand during a period, datacenters usually overly provision and keep more running servers than what is needed to satisfy the peak load. In our experiment, we assume that the datacenter has the complete workload information ahead of time and provisions exactly to satisfy the peak load. Using such benchmark gives us a conservative estimate of the cost saving from our algorithms.

Sever operation cost: The server operation cost is determined by unit-time energy cost P and on-off costs β_{on} and β_{off}. In the experiment, we assume that a server consumes one unit energy for per unit time, i.e., $P = 1, \forall x$. We set $\beta_{off} + \beta_{on} = 6$, i.e., the cost of turning a server off and on once is equal to that of running it for six units of time [1]. Under this setting, the critical interval is $\Delta = (\beta_{off} + \beta_{on})/P = 6$ units of time.

6.2   Performance of the Proposed Online Algorithms

We have characterized in the theorems in Chaps. 4–5 the competitive ratios of **PrOnSCP** and **RPrOnSCP** as the look-ahead window size, i.e., $\alpha_s \Delta$, increases, where $\Delta = P_{min} d_{min} \frac{1}{\beta_s}$. The resulting competitive ratios, i.e., $2 - \alpha_s$ and $e/(e - 1 + \alpha_s)$, already appealing, are for the worst-case scenarios. In practice, the actual performance can be even better.

In our first experiment, we study the performance of **PrOnSCP** and **RPrOnSCP** using real-world traces. The cost reductions are shown in Fig. 6.2. The cost reduction curves are obtained by comparing the power cost incurred by the offline algorithm, **PrOnSCP**, **RPrOnSCP**, the **LCP**(ω) algorithm [1] and the **DELAYEDOFF** algorithm [2] to the cost benchmark. The vertical axis indicates the cost reduction and the horizontal axis indicates the size of look-ahead window varying from 0 to 10 units of time.

The curves of normalized numbers of servers(normalized by the maximal number of servers in [0, T]) given by **PrOnSCP/RPrOnSCP** are shown in Fig. 6.3. In order to show

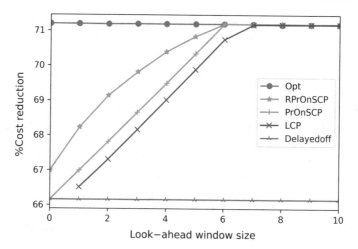

Fig. 6.2 Impact of future information

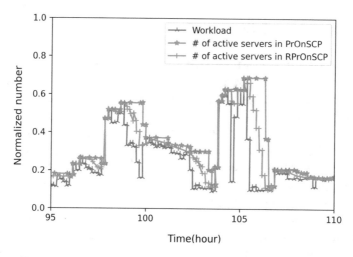

Fig. 6.3 Normalized numbers of jobs and active servers given by **PrOnSCP/RPrOnSCP** in a period

more details, only the numbers of active servers in a period (from hour 95 to hour 110) are plotted. Figure 6.3 indicates that the numbers of active servers used by **PrOnSCP** and **RPrOnSCP** decrease when workload is low and increase when workload is high and similar patterns are observed in the rest period. Those curves actually match the intuitive strategy that in order to save energy we should turn on just enough servers to meet the demand. It can also be seen that **RPrOnSCP** is more aggressive in turning servers off as compared to **PrOnSCP**, which for this workload trace leads to the observation that **RPrOnSCP** reduces more operating cost than **PrOnSCP**.

For this workload, **PrOnSCP**, **RPrOnSCP**, **LCP**(ω) and **DELAYEDOFF** achieve substantial cost reduction as compared to the benchmark. In particular, the cost reductions of **PrOnSCP** and **RPrOnSCP** are beyond 66% even when no future workload information is available. **LCP**(ω) starts to perform when the look-ahead window size is one. This is because we run **LCP**(w) under a discrete-time setting and the workload information for the current slot is only available after all jobs in this slot have arrived. Meanwhile, **PrOnSCP**, **RPrOnSCP**, and **DELAYEDOFF** are running under a continuous-time setting, where jobs arriving at any moments are served immediately.

The cost reductions of **PrOnSCP** and **RPrOnSCP** grow linearly as the look-ahead window increases, and reaching optimal when the look-ahead window size reaches Δ. These observations match what our theorem predicts. Meanwhile, **LCP**(ω) has not yet reach the optimal performance when the look-ahead window size reaches the critical value Δ. **DELAYEDOFF** has the same performance for all look-ahead window sizes since it does not exploit future workload information.

6.3   Impact of Prediction Error

Previous experiments show that **PrOnSCP**, **RPrOnSCP** and **LCP**(ω) have better performance if accurate future workload is available. However, there are always prediction errors in practice. Therefore, it is important to evaluate the performance of the algorithms in the present of prediction error.

To achieve this goal, we evaluate **PrOnSCP** and **RPrOnSCP** with look-ahead window size of 2 and 4 units of time. Zero-mean Gaussian prediction error is added to each unit-time workload in the look-ahead window, with its standard deviation grows from 0 to 50% of the corresponding actual workload. In practice, prediction error tends to be small [5]; thus we are essentially stress-testing the algorithms.

We average 100 runs for each algorithm and show the results in Fig. 6.4, where the vertical axis represents the cost reduction as compared to the benchmark.

On one hand, we observe all algorithms are fairly robust to prediction errors. On the other hand, all algorithms achieve better performance with a look-ahead window size 4 than size 2. This indicates more future workload information, even inaccurate, is still useful in boosting the performance.

6.4   Impact of Peak-to-Mean Ratio (PMR)

Intuitively, comparing to static provisioning, dynamic provisioning can save more power when the datacenter trace has large PMR. Our experiments confirm this intuition which is also observed in other work [1, 4]. Similar to [1], we generate the workload from the MSR traces by scaling $a\,(t)$ as $\overline{a\,(t)} = Ka^{\gamma}\,(t)$, and adjusting γ and K to keep the mean constant. We run the offline algorithm, **PrOnSCP**, **RPrOnSCP**, **LCP**(ω) and **DELAYEDOFF** using

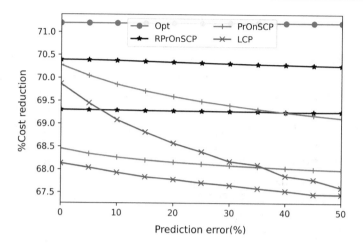

Fig. 6.4 Impact of prediction error

workloads with different PMRs ranging from 2 to 10, with look-ahead window size of one unit time. The results are shown in Fig. 6.5.

As seen, energy saving increases form about 40% at PRM = 2, which is common in large datacenters, to large values for the higher PMRs that is common in small to medium sized datacenters. Similar results are observed for different look-ahead window sizes.

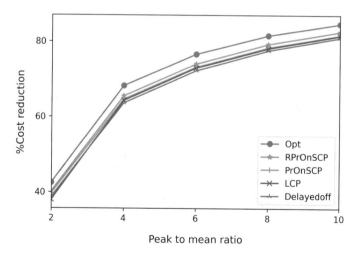

Fig. 6.5 Impact of PMR

6.5   Discussion

Note that **PrOnSCP** and **RPrOnSCP** have competitive ratios $2 - \alpha_s$ and $e/(e - 1 + \alpha_s)$, which improve as future information is available. This is in contrast to **LCP**(w), whose best known competitive ratio is 3 and, regardless of how much future information is available, there are instances with performance arbitrarily close to the ratio. Figure 6.2 shows that **PrOnSCP/RPrOnSCP** perform slightly better than **LCP**(ω), partially because they need not work in discrete time. These performance gains of **PrOnSCP/RPrOnSCP** over **LCP**(ω) and **DELAYEDOFF** shown in Fig. 6.2, when multiplying the large amount of energy consumed by the datacenters every year, correspond to non-negligible energy cost saving. Moreover, the sleep management in **PrOnSCP/RPrOnSCP** are decentralized, which makes them very much easier to implement; while the **LCP**(w) is inherently centralized, since it requires the solution of a convex program at each time.

Although in this example, **DELAYEDOFF** performs close to the optimal, there are very natural cases in which it can be almost a factor of two more expensive than **PrOnSCP/RPrOnSCP**. The value of Δ is approximately one hour [1], and it is common for workloads to have a periodic structure with a period one hour. In this case, it is possible that **DELAYEDOFF** always turns machines off just before they are needed again. If the workload can be predicted an hour into the future, then **PrOnSCP/RPrOnSCP** can guarantee optimal performance in this case. **DELAYEDOFF** also does not exploit randomness to improve performance like **RPrOnSCP** does.

References

1. M. Lin, A. Wierman, L. Andrew, and E. Thereska. Dynamic right-sizing for power-proportional data centers. In *Proc. IEEE INFOCOM*, pages 1098–1106, 2011.
2. A. Gandhi, V. Gupta, M. Harchol-Balter, and M. Kozuch. Optimality analysis of energy-performance trade-off for server farm management. *Performance Evaluation*, 2010.
3. D. Narayanan, A. Donnelly, and A. Rowstron. Write off-loading: Practical power management for enterprise storage. *ACM Transactions on Storage (TOS)*, 4(3): 10, 2008.
4. A. Krioukov, P. Mohan, S. Alspaugh, L. Keys, D. Culler, and R. Katz. Napsac: design and implementation of a power-proportional web cluster. *ACM SIGCOMM Computer Communication Review*, 41(1): 102–108, 2011.
5. D. Kusic, J. Kephart, J. Hanson, N. Kandasamy, and G. Jiang. Power and performance management of virtualized computing environments via lookahead control. *Cluster Computing*, 12(1): 1–15, 2009.

Conclusion and Extensions

7

Dynamic provisioning is an effective technique for reducing server energy consumption in datacenters, by turning off unnecessary servers to save energy. In this monograph, we design online dynamic provisioning algorithms with zero or partial future workload information available.

We reveal an elegant structure of the off-line dynamic provisioning problem, under the cost model that a running server consumes a fixed amount of energy per unit time. Exploiting such structure, we show its optimal solution can be achieved by the datacenter adopting a simple last-empty-server-first job-dispatching strategy and each server independently solving a classic ski-rental problem.

We build upon this architectural insight to design two new decentralized online algorithms. One is deterministic with competitive ratio $2 - \alpha_s$, where $0 \leq \alpha_s \leq 1$ is the fraction of the full-size look-ahead window in which future workload information is available. The size of the full-size look-ahead window is determined by the wear-and-tear cost and the unit-time energy cost of running a single server. The another is randomized with a competitive ratio $\frac{e}{(e-1+\alpha_s)}$. The ratios $2 - \alpha_s$ and $\frac{e}{(e-1+\alpha_s)}$ are the best competitive ratios for any deterministic and randomized online algorithms under last-empty-server-first job-dispatching strategy. Note that the problem we study in this monograph is similar to that studied in [1]. The difference is that we optimize a linear cost function over integer variables, while Lin et al. in [1] minimize a convex cost function over continuous variables (by relaxing the integer constraints). This monograph and [1] obtain different online algorithms with different competitive ratios for the two different formulations, respectively.

Our algorithms are simple and easy to implement. Simulations using real-world traces show that our algorithms can achieve close-to-optimal energy-saving performance, and are robust to future-workload prediction errors. These results suggest that it is possible to reduce server energy consumption significantly with zero or only partial future workload information.

© The Author(s), under exclusive license to Springer Nature Switzerland AG 2022
M. Chen and S. C.-K. Chau, *Online Capacity Provisioning for Energy-Efficient Datacenters*, Synthesis Lectures on Learning, Networks, and Algorithms,
https://doi.org/10.1007/978-3-031-11549-3_7

Our results lead to a fundamental observation that under the cost model that a running server consumes a fixed amount of energy per unit time, future workload information beyond the the the full-size look-ahead window will not improve the dynamic provisioning performance. We also believe utilizing future input information is a new and important design degree freedom for online algorithms.

7.1   Extensions for Practical Datacenters

In this section, we describe several extensions of our algorithms for practical datacenters considering diverse real-world settings.

First, we can extend the online algorithms for the case that servers need setup time T_s but the load satisfies $a(\tau) \le (1 + \gamma)a(t)$ for all $\tau \in [t, t + T_s]$. These algorithms have competitive ratios $(2 - \alpha_s)(1 + \gamma) + 2\gamma$ and $\frac{e}{e-1+\alpha_s}(1 + \gamma) + 2\gamma$. We now describe a centralized algorithm EXT that provides a bounded CR in the case where $a(\tau) \le (1 + \gamma)a(t)$ for all $\tau \in [t, t + T_s]$, i.e., workload increases at most by a factor of $(1 + \gamma)$ in any interval of length T_s. We expect γ to be small. In this model, servers can be in three states: ON, BOOT, OFF. Only servers in state ON can serve jobs, but servers in states ON and BOOT both consume power P per unit time. An OFF server "turns on" when it enters state BOOT; T_s later it will become ON. A server in any state can immediately be turned OFF.

Algorithm EXT for Cases with Setup Time:
Each server:

- Behaves as for **PrOnSCP** or **RPrOnSCP**, but when its timer expires, it does not turn off but sends a message M to manager.

Centralized Controller:
Keeps track of the set X (of size x) of "active" servers, i.e., those that have not sent M since being allocated a job. It responds to two types of events as follows:

- Job arrival: If X contains an idle server, the job is sent to a server in X using last-empty-server-first in **PrOnSCP** or least-idle in **RPrOnSCP**. Otherwise it is sent to another ON server. Additional servers will be turned on so that the total number of ON and BOOT servers are $\lfloor x(1 + \gamma) \rfloor + 1$.
- Message M from server: All but $\lceil x(1 + \gamma) \rceil + 1$ servers will be turned OFF. BOOT servers are turned off first, in decreasing order of how recently they were turned on. No active servers are turned off.

The following result establishes the validity and performance guarantees of EXT.

Corollary 7.1.1 *If there are $\lceil a(0)(1+\gamma)\rceil + 1$ ON servers at time 0, then under EXT, the number of ON servers at time t is at least $a(t)$. Let $a_{\min} = \min_{t \in [0,T]} a(t)$. The competitive ratio of EXT on instances with discrete arrival instants is $(2 - \alpha_s)(1+\gamma) + 2/a_{\min}$ if servers use **PrOnSCP**, or $\frac{e}{e-1+\alpha_s}(1+\gamma) + 2/a_{\min}$ if servers use **RPrOnSCP**. These are bounded above by $(2 - \alpha_s)(1+\gamma) + 2\gamma$ and $\frac{e}{e-1+\alpha_s}(1+\gamma) + 2\gamma$.*

Remarks: (i) The competitive ratio of EXT is linearly proportional to γ. (ii) Since the minimal workload $a_{\min}$ in large data centers is normally much larger than that in small ones, whence EXT is usually more beneficial for large data center. (iii) In EXT, we adopt over-provisioning to combat the problem that servers need setup time T_s; it would be interesting to know if there exist other approaches to handle this problem. EXT can not achieve competitive ratio 1. Hence, it is also good to know how to better utilize future information.

This work can also be extended in many important directions. In the elephant model considered here, each server could only serve one job at a time. Cloud computing datacenters typically run multiple VMs on each physical machine. One particular motivation is to pack together jobs with complementary resource requirements, such as placing a CPU-intensive and a memory-intensive VM on the same server. In this scenario, minimizing the total power cost is a dynamic bin-packing problem which is NP-hard. (It contains classic bin packing as a special case.). The analysis of dynamic bin-packing problem is entirely different and it would be interesting to look at it in the future. Even in the simplest case that each server can host an arbitrary combination of m VMs, the problem is significantly different; it is no longer the case that the optimal performance can be obtained by a non-clairvoyant algorithm without VM migration, and indeed such algorithms are at best m- competitive. A related extension would be to consider the fact that VMs may have time-varying resource requirements. When utilizing future workload information, we assume that the future information is accurate in the look-ahead window in our algorithm design and competitive analysis, and we study in experiments the performance of the proposed algorithms when the future information is not perfectly known. An interesting and important future direction is to design competitive online algorithms that can utilize inaccurate future input information. Such algorithms will be very attractive in practice, where prediction of the future input information often comes with errors.

Another important direction would be to extend these results to the general case of heterogeneous servers or multiple geographically separated datacenters [2–4]. It would be useful to extend the insight from this monograph to heterogeneous cases.

7.2 Online Algorithms using Machine Learning

The advent of machine learning enables effective prediction algorithms that can attain good prediction accuracy from a sufficient set of training data. Hence, recent studies [5, 6] have incorporated predictions in online algorithms to improve the competitive ratios, when given accurate predictions. It is worth mentioning that predictions are not always perfect in practice. It is critical to characterize and limit the degradation of imperfect predictions to prediction-aware online algorithms. The notions of consistency (considering accurate predictions) and robustness (considering inaccurate predictions) can shed light on the effectiveness of prediction-aware online algorithms.

Traditional competitive online algorithms do not rely on any predictions. Hence, they always act on the consideration of worst-case future events (but often pessimistically). Recently, prediction-aware online algorithms have been studied in the literature [7, 8]. In these algorithms, there is normally a hyperparameter $\omega \in [0, 1]$ that characterizes the extent of predictions to be harnessed in an online algorithm. In our convention,[1] we use $\omega = 0$ to represent the traditional competitive online algorithm scenario without prediction, whereas $\omega = 1$ to represent the ideal scenario of utilizing full predictions to achieve an offline optimal solution based on the assumption of accurate predictions. An online algorithm is said to have a consistently competitive ratio $\alpha(\omega)$ (or $\alpha(\omega)$-consistent), if its competitive ratio is bounded by a function $\alpha(\omega)$ dependent on ω, assuming that the predictions utilized in the algorithm are entirely accurate. Also, an online algorithm is said to have a robustly competitive ratio $\beta(\omega)$ (or $\beta(\omega)$-robust), if its competitive ratio is bounded by a function $\beta(\omega)$, regardless the predictions utilized in the algorithm are accurate or not. There is a trade-off between the consistency of harnessing accurate predictions and the robustness against falling for inaccurate predictions. Hence, increasing ω will decrease $\alpha(\omega)$, but also will increase $\beta(\omega)$.

Particularly, we consider a prediction-aware online algorithm that is associated with a predictor. In the rent-or-buy problem, we define a critical time window by $[t, t + B]$. Evidently, at the current timeslot t, the predictor needs to consider no more than $[t, t + B]$ for the rent-or-buy decision. The predictor is represented by a time-dependent function $\Pi_t(\omega)$, which only produces predictions within a sliding prediction window $[t, t + \omega B]$. The predictor can be invoked at any timeslot t to aid the online decisions. Note that $\Pi_t(\omega)$ may not produce the same predictions at all t. It is important to capture the temporal variations in predictions. For instance, $\Pi_t(\omega)$ predicts the end of usage duration in $[t, t + \omega B]$. When approaching the actual end of usage duration, $\Pi_t(\omega)$ will be able to make increasingly accurate predictions, possibly based on further revealed information of usage pattern. Namely, $\Pi_t(\omega)$ is more accurate than $\Pi_{t'}(\omega)$, for $t' < t$. In particular, the predictor precisely knows the end of usage, when $t = D$.

We next describe our recent online algorithmic results in the ski-rental problem with consistency and robustness guarantees.

---

[1] We adopt a slightly different convention than [7, 8]. In their papers, they use hyperparameter $\lambda = 1$ to represent the scenario without prediction, and $\lambda = 0$ to represent the ideal scenario with full predictions.

Theorem 7.2.1 *There is a $(2 - \omega)$-consistent and $(\frac{2-\omega}{1-\omega})$-robust deterministic online algorithm, and a a $\left(\frac{e-\omega}{e-1}\right)$-consistent and $\left(\frac{e-\omega}{(e-1)(1-\omega)}\right)$-robust randomized online algorithm in expectation for the ski-rental problem.*

7.2.1 Deterministic Online Algorithm

First, we present a deterministic sliding prediction window based algorithm (dSPW), which is a simple online algorithm considering sliding prediction window $[t, t + \omega B]$.

Algorithm 7.1 Deterministic online algorithm **dSPW**(ω)

Input: Predictor $\Pi_t(\omega)$
Output: Renting or buying
Initialization: t_0 is the starting time of renting
1: $t \leftarrow t_0$
2: **while** resource is needed at time t **do**
3:    **if** $\Pi_t(\omega) \neq$ Null **then**
4:       $\hat{x} \leftarrow \Pi_t(\omega)$
5:       $\hat{k} \leftarrow \hat{x} - t_0$ // *Predict the number of days of skiing $\hat{k}$*
6:    **else**
7:       $\hat{k} \leftarrow t + \omega B - t_0$ // *Assume $\hat{x} \geq t + \omega B$*
8:    **end if**
9:    **if** $\hat{k} \geq B$ or $t - t_0 \geq B$ **then**
10:      Buying: $x_t \leftarrow 0$
11:      Exit
12:    **else**
13:      Continue renting: $x_t \leftarrow 1$ // *Stay in the while loop*
14:    **end if**
15:    $t \leftarrow t + \Delta t$ // *Make another decision after Δt*
16: **end while**
17: Return x_t

Lemma 7.2.2 *dSPW is a $(2 - \omega)$-consistent and $(\frac{2-\omega}{1-\omega})$-robust online algorithm.*

Proof Note that the cost of the offline optimal solution is

$$\text{Cost(Opt)} = \begin{cases} k, & \text{if } k < B, \\ B, & \text{otherwise} \end{cases}$$

Without loss of generality, we assume $t_0 = 0$.

First, we show the consistency by assuming that $\Pi_t(\omega)$ is always accurate. We consider three cases: (1) $k < \omega B$, (2) $\omega B \leq k < B$, and (3) $k \geq B$.

- Case (1): When $k < \omega B$, **dSPW** is always able to predict k in $[t, t + \omega B]$. Hence, Cost(**dSPW**) $= k$.
- Case (2): When $\omega B \leq k < B$, **dSPW** rents for a duration $k - \omega B$, and then utilizes $\Pi_{k-\omega B}(\omega)$ to obtain $k(< B)$, and rents until the end of usage duration. Hence, Cost(**dSPW**) $= k$.
- Case (3): When $k \geq B$, **dSPW** rents for a duration at most $B - \omega B$, and then utilizes $\Pi_{k-\omega B}(\omega)$ to infer that $k \geq B$ and decides to buy. Hence, Cost(**dSPW**) $\leq B - \omega B + B = B(2 - \omega)$.

In cases (1)-(2), Cost(Opt) $= k$. In case (3), Cost(Opt) $= B$. Hence, the consistently competitive ratio is

$$\frac{\text{Cost}(\textbf{dSPW})}{\text{Cost}(\text{Opt})} \leq 2 - \omega$$

Next, we show the robustness by considering that $\Pi_t(\omega)$ may be misleading. We consider three cases: (1) $k < B - \omega B$, (2) $B - \omega B \leq k < B$, and (3) $k \geq B$.

- Case (1): Since $k < B - \omega B \Rightarrow k + \omega B < B$, **dSPW** will not be triggered to buy before $t = k$, despite any inaccurate predictions. Hence, Cost(**dSPW**) $= k$.
- Case (2): Since $B - \omega B \leq k \Rightarrow k + \omega B \geq B$, **dSPW** will be possibly misled by inaccurate predictions to buy at $t \leq k$. When $t > k$, the decision maker can see the end of usage regardless of the predictor. Hence, Cost(**dSPW**) $\leq k + B$.
- Case (3): When $k \geq B$, **dSPW** will be possibly misled by inaccurate predictions to continue renting. However, **dSPW** will only rent for a duration at most B, and then buy. Hence, Cost(**dSPW**) $\leq 2B$.

In cases (1)-(2), Cost(Opt) $= k$. In case (3), Cost(Opt) $= B$. Hence, case (2) is the worst case, and the robustly competitive ratio is

$$\frac{\text{Cost}(\textbf{dSPW})}{\text{Cost}(\text{Opt})} \leq \frac{k + B}{k} = 1 + \frac{B}{k} \leq 1 + \frac{B}{B - \omega B} = \frac{2 - \omega}{1 - \omega} \qquad \square$$

7.2.2   Randomized Online Algorithm

Next, we present a randomized sliding prediction window based algorithm (**rSPW**). The basic idea of **rSPW** is to reduce the break-even threshold probabilistically, and trigger buying earlier. We define a probability density function $f(\lambda)$:

$$f(\lambda) \triangleq \begin{cases} \dfrac{e^{\frac{\lambda}{B(1-\omega)}}}{B(1-\omega)(e-1)}, & \text{if } 0 \le \lambda \le (1-\omega)B, \\ 0, & \text{otherwise} \end{cases} \tag{7.1}$$

Algorithm 7.2 Randomized online algorithm **rSPW**(ω)

Input: Predictor $\Pi_t(\omega)$
Output: Renting or buying
Initialization: t_0 is the starting time of renting
1: $t \leftarrow t_0$
2: Randomly generate λ according to $f(\lambda)$
3: **while** resource is needed at time t **do**
4:    **if** $\Pi_t(\omega) \ne$ Null **then**
5:        $\hat{x} \leftarrow \Pi_t(\omega)$
6:        $\hat{k} \leftarrow \hat{x} - t_0$
7:    **else**
8:        $\hat{k} \leftarrow t + \omega B - t_0$
9:    **end if**
10:   **if** $\hat{k} \ge \lambda + \omega B$ or $t - t_0 \ge \lambda + \omega B$ **then**
11:       Buying: $x_t \leftarrow 0$
12:       Exit
13:   **else**
14:       Continue renting: $x_t \leftarrow 1$
15:   **end if**
16:   $t \leftarrow t + \Delta t$
17: **end while**
18: Return x_t

Lemma 7.2.3 *rSPW is a $\left(\frac{e-\omega}{e-1}\right)$-consistent and $\left(\frac{e-\omega}{(e-1)(1-\omega)}\right)$-robust online algorithm in expectation.*

Proof Without loss of generality, we assume $t_0 = 0$.

First, we show the consistency by assuming that $\Pi_t(\omega)$ is always accurate. Next, we compute $\mathbb{E}[\text{Cost}(\textbf{rSPW})]$. Without loss of generality, we assume $t_0 = 0$. We consider three cases: (1) $k < \omega B$, (2) $\omega B \le k < B$, and (3) $k \ge B$.

- Case (1): When $k < \omega B$, **rSPW** is always able to predict k in $[t, t + \omega B]$. Hence $\mathbb{E}[\text{Cost}(\textbf{rSPW})] = k$.
- Case (2): When $\omega B \le k < B$, **rSPW** may or may not be able to predict k, conditional on the value of random λ. When $k \le \lambda + \omega B$, then **rSPW** is able to predict k in $[t + \lambda, t + \lambda + \omega B]$. Hence, $\mathbb{E}[\text{Cost}(\textbf{rSPW})] = k$.

However, when $\lambda + \omega B < k$, then **rSPW** is not able to predict k in $[t + \lambda, t + \lambda + \omega B]$. Instead, **rSPW** buys at the end of λ period. So, $\text{Cost}(\textbf{rSPW}) = \lambda + B$.

In this case, we compute $\mathbb{E}[\text{Cost}(\textbf{rSPW})]$ as follows:

$$\mathbb{E}[\text{Cost}(\textbf{rSPW})]$$
$$= \int_{k-\omega B}^{B-\omega B} kf(\lambda)\mathrm{d}\lambda + \int_{0}^{k-\omega B} (\lambda + B)f(\lambda)\mathrm{d}\lambda$$

Note that

$$\int_{k-\omega B}^{B-\omega B} kf(\lambda)\mathrm{d}\lambda = \int_{k-\omega B}^{B-\omega B} \frac{ke^{\frac{\lambda}{B(1-\omega)}}}{B(1-\omega)(e-1)}\mathrm{d}\lambda$$
$$= \frac{ke - ke^{\frac{\omega B-k}{B(1-\omega)}}}{e-1}$$

$$\int_{0}^{k-\omega B} (\lambda + B)f(\lambda)\mathrm{d}\lambda = \int_{0}^{k-\omega B} \frac{(\lambda + B)e^{\frac{\lambda}{B(1-\omega)}}}{B(1-\omega)(e-1)}\mathrm{d}\lambda$$
$$= \frac{ke^{\frac{\omega B-k}{B(1-\omega)}} - \omega B}{e-1}$$

Hence,

$$\mathbb{E}[\text{Cost}(\textbf{rSPW})] = \frac{ke - \omega B}{e-1} \le \frac{k(e-\omega)}{e-1}$$

- Case (3): When $k \ge B$, then **rSPW** is not able to predict k in $[t + \lambda, t + \lambda + \omega B]$. Instead, **rSPW** buys at the end of λ period. Hence, $\text{Cost}(\textbf{rSPW}) = \lambda + B$.

$$\mathbb{E}[\text{Cost}(\textbf{rSPW})] = \int_{0}^{B-\omega B} (\lambda + B)f(\lambda)\mathrm{d}\lambda \qquad (7.2)$$
$$= \frac{B(e-\omega)}{e-1} \qquad (7.3)$$

In cases (1)-(2), $\text{Cost}(\text{Opt}) = k$. In case (3), $\text{Cost}(\text{Opt}) = B$. Hence, the expected consistently competitive ratio is

$$\frac{\mathbb{E}[\text{Cost}(\textbf{rSPW})]}{\text{Cost}(\text{Opt})} \le \frac{e-\omega}{e-1}$$

Next, we show the robustness by considering that $\Pi_t(\omega)$ may be misleading. We consider three cases: (1) $k < B - \omega B$, (2) $B - \omega B \le k < B$, and (3) $k \ge B$.

- Case (1): When $\lambda < k$, **rSPW** is renting in $[t, t + \lambda]$ and may be misled by inaccurate predictions to buy at the end of duration λ. However, when $\lambda \ge k$, **rSPW** does not rely on any (inaccurate) prediction, and it only rents for a duration at most k. Hence,

$$\mathbb{E}[\text{Cost}(\textbf{rSPW})]$$

$$= \int_0^k (\lambda + B)f(\lambda)\mathrm{d}\lambda + \int_k^{B-\omega B} kf(\lambda)\mathrm{d}\lambda$$

$$= \frac{ke + \omega B e^{\frac{k}{B-\omega B}} - \omega B}{e - 1}$$

- Case (2): Since $B - \omega B \leq k \Rightarrow k + \omega B \geq B$, **rSPW** may be misled by inaccurate prediction to buy at $t = \lambda$. Hence,

$$\mathbb{E}[\text{Cost}(\textbf{rSPW})] = \int_0^{B-\omega B} (\lambda + B)f(\lambda)\mathrm{d}\lambda$$

$$= \frac{B(e - \omega)}{e - 1}$$

- Case (3): When $k \geq B$, we obtain

$$\mathbb{E}[\text{Cost}(\textbf{rSPW})] = \int_0^{B-\omega B} (\lambda + B)f(\lambda)\mathrm{d}\lambda$$

$$= \frac{B(e - \omega)}{e - 1}$$

In cases (1)-(2), $\text{Cost}(\text{Opt}) = k$. In case (3), $\text{Cost}(\text{Opt}) = B$. In case (1), we obtain

$$\frac{\mathbb{E}[\text{Cost}(\textbf{rSPW})]}{\text{Cost}(\text{Opt})} = h(k) \triangleq \frac{ke + \omega B e^{\frac{k}{B-\omega B}} - \omega B}{k(e - 1)}$$

We note that $h'(k) \geq 0$ and $h''(k) \geq 0$ for $k \in [0, B - \omega B]$. Hence, $h(k) \leq h(B - \omega B) = \frac{e-\omega}{(1-\omega)(e-1)}$. In case (2), $\frac{\mathbb{E}[\text{Cost}(\textbf{rSPW})]}{\text{Cost}(\text{Opt})}$ is maximized, when $k = B - \omega B$:

$$\frac{\mathbb{E}[\text{Cost}(\textbf{rSPW})]}{\text{Cost}(\text{Opt})} \leq \frac{B(e - \omega)}{k(e - 1)} \leq \frac{B(e - \omega)}{(B - \omega B)(e - 1)}$$

$$= \frac{e - \omega}{(1 - \omega)(e - 1)}$$

Therefore, we show that **rSPW** is a $\left(\frac{e-\omega}{e-1}\right)$-consistent and $\left(\frac{e-\omega}{(e-1)(1-\omega)}\right)$-robust online algorithm. $\qquad \square$

Fig. 7.1 Comparison with
robust online algorithms

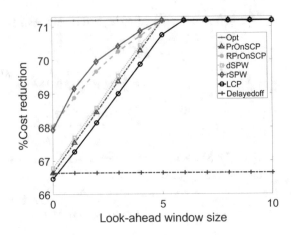

7.2.3   Empirical Evaluation

Next, we apply these online algorithms with robustness guarantees to online capacity provisioning for energy-efficient datacenters (**dSPW** and **rSPW**). We set the the look-ahead window size as $\alpha_s \Delta$, where $\Delta \triangleq P_{min} d_{min} \frac{1}{\beta_s}$.

We present some empirical studies of these online algorithms with **PrOnSCP** and **RPrOnSCP**. In Fig. 7.1, we observe that the online algorithms with robustness guarantees slightly outperform **PrOnSCP** and **RPrOnSCP** to a certain extent.

7.3   Other Online Scheduling Problems

In this monograph, we studied the online capacity provisioning problem, with a connection to ski-rental problem. However, there are a wide range of other online scheduling problems. We provide a brief discussion about these problems.

There is an online generation scheduling problem in microgrid [9], which considers a typical scenario where a microgrid orchestrates different energy generation sources to minimize cost. The microgrid must satisfy both electricity and heat demands simultaneously, while meeting operational constraints of the electric power system. The generation scheduling problem aims to control the types of power generators with respect to different demands to minimize the generation cost. Different types of power generators will have different generation constraints, for example, having a minimum period of time for starting-up or shutting-down. Hence, scheduling the generation in the presence of uncertain future demands will require a sophisticated online algorithm. Although there are similarities with the online capacity provisioning problem, generation dispatching problem has some unique characteristics of state transitions, which will give rise to long-time dependency in the online decisions. Generation dispatching problem is usually studied under the general formulation of Metric Task Systems (MTS) problem [10].

References

1. M. Lin, A. Wierman, L. Andrew, and E. Thereska. Dynamic right-sizing for power-proportional data centers. In *Proc. IEEE INFOCOM*, pages 1098–1106, 2011.
2. M. Lin, Z. Liu, A. Wierman, and L. L. H. Andrew. Online algorithms for geographical load balancing. In *Proc. Int. Green Computing Conf.*, 2012.
3. R. Nathuji, C. Isci, and E. Gorbatov. Exploiting platform heterogeneity for power efficient data centers. In *Autonomic Computing, 2007. ICAC'07. Fourth International Conference on*, pages 5–5. IEEE, 2007.
4. T. Heath, B. Diniz, E. Carrera, W. Meira Jr, and R. Bianchini. Energy conservation in heterogeneous server clusters. In *Proceedings of the tenth ACM SIGPLAN symposium on Principles and practice of parallel programming*, pages 186–195. ACM, 2005.
5. M. Mitzenmacher and S. Vassilvitskii. Algorithms with predictions. In T. Roughgarden, editor, *Beyond the Worst-Case Analysis of Algorithms*. Cambridge University Press, 2020.
6. T. Lykouris and S. Vassilvitskii. Competitive caching with machine learned advice. In *Proc. International Conf. on Machine Learning (ICML)*, 2018.
7. R. Kumar, M. Purohit, and Z. Svitkina. Improving online algorithms via ML predictions. In *Proc. Annual Conf. on Neural Information Processing Systems (NeurIPS)*, 2018.
8. S. Gollapudi and D. Panigrahi. Online algorithms for rent-or-buy with expert advice. In *Proc. International Conf. on Machine Learning (ICML)*, 2019.
9. L. Lu, J. Tu, C.-K. Chau, M. Chen, and X. Lin. Online energy generation scheduling for microgrids with intermittent energy sources and co-generation. In *ACM SIGMETRICS*, 2013.
10. A. Borodin and R. El-Yaniv. *Online Computation and Competitive Analysis*. Cambridge University Press, 2005.

Appendix

Theorem A.1 *Consider arbitrary convex and increasing functions $g_t(x(t), a(t))$, $t \in [1, T]$. Under the online setting where at time t, $a(\tau) \in [0, 1]$, $p(\tau)$, and $g_\tau(x(\tau), a(\tau))$, for all $\tau \in [0, t + \omega]$, are available, **PrOnSCP** (Algorithm 4.4) achieves the best possible competitive ratio $2 - \alpha_s$ among all deterministic online algorithms for **SCP**. Here*

$$\alpha_s \triangleq \min\left(1, \omega \cdot P_{\min} d_{\min} \frac{1}{\beta_s}\right) \in [0, 1],$$

is a "normalized" prediction window duration and

$$d_{\min} \triangleq \min_{t \in [1,T]} \left\{ g_t(1, 0) - g_t(0, 0) \right\}. \tag{A.1}$$

Lemma A.1 *PrOnSCP is $(2 - \alpha_s)$-competitive for SCP.*

Proof We compare our prediction-aware online algorithm **PrOnSCP** and the offline optimal algorithm **OffSCP** for problem **SCP** and prove the competitive ratio.

Let x^{on} and $\bar{x}$ be the solutions obtained by **PrOnSCP** and **OffSCP** for problem **SCP**, respectively.

It is easy to see that during I_s and I_2, **PrOnSCP** and **OffSCP** have the same actions. Since the adversary can choose the T to be large enough, we can omit the cost incurred during I_e when doing competitive analysis. Thus, we only need to consider the cost incurred by the **PrOnSCP** and **OffSCP** during each I_1. Notice that at the beginning of an I_2, both algorithms may incur switching cost. However, there must be an I_1 before an I_2. So this switching cost will be taken into account when we analyze the cost incurred during I_1. More formally, for a certain I_1, denoted as $[t_1, t_2]$,

© The Editor(s) (if applicable) and The Author(s), under exclusive license to Springer Nature Switzerland AG 2022
M. Chen and S. C.-K. Chau, *Online Capacity Provisioning for Energy-Efficient Datacenters*, Synthesis Lectures on Learning, Networks, and Algorithms,
https://doi.org/10.1007/978-3-031-11549-3

$\text{Cost}_{I_1}(x)$

$$= \sum_{t=t_1}^{t_2} p(t) \cdot g_t(x(t), 0) + \beta_s \sum_{t=t_1}^{t_2+1} [x(t) - x(t-1)]^+$$

$$= \sum_{t=t_1}^{t_2} p(t) d_t x(t) + \sum_{t=t_1}^{t_2} p(t) g_t(0, 0) + \beta_s \sum_{t=t_1}^{t_2+1} [x(t) - x(t-1)]^+. \quad (A.2)$$

where $a(t) \in [0, 1]$, and $d_t \triangleq g_t(1, 0) - g_t(0, 0)$. We let the constant cost $c = \sum_{t=t_1}^{t_2} p(t) g_t(0, 0)$ which is independent of $x(t)$.

PrOnSCP performs as follows: it accumulates an "idling cost" and when it reaches β_s, it turns off the server; otherwise, it keeps the server idle. Specifically, at time t, if there exists $\tau \in [t, t+w]$ such that the idling cost till τ is at least β_s, it turns off the server; otherwise, it keeps it idle. We distinguish two cases:

Case 1: $w \geq \beta_s/(d_{\min} P_{\min})$. In this case, **PrOnSCP** performs the same as **OffSCP**. Because of the following reasons:

If $\sum_{t \in I_1} d_t p(t) \geq \beta_s$, **OffSCP** turns off the server at the beginning of the I_1, i.e., at t_1. Since $w \geq \beta_s/(d_{\min} P_{\min})$ and $d_t \geq d_{\min}$ according to Eq. (4.4), at t_1 **PrOnSCP** can find a $\tau \in [t_1, t_1 + w]$ such that the idling cost till τ is at least β_s, as a consequence of which it also turns off the server at the beginning of the I_1. Both algorithms turn on the server at the beginning of the following I_2. Thus, we obtain

$$\text{Cost}_{I_1}(x^{on}) = \text{Cost}_{I_1}(\bar{x}) = \beta_s + c. \quad (A:3)$$

If $\sum_{t \in I_1} d_t p(t) < \beta_s$, **OffSCP** keeps the server idling during the whole I_1. **PrOnSCP** finds that the accumulate idling cost till the end of the I_1 will not reach β_s, so it also keeps the server idling during the whole I_1. Thus, we have

$$\text{Cost}_{I_1}(x^{on}) = \text{Cost}_{I_1}(\bar{x}) = \sum_{t \in I_1} d_t p(t) + c. \quad (A.4)$$

Case 2: $w < \beta_s/(d_{\min} P_{\min})$. In this case, to beat **PrOnSCP**, the adversary will choose $p(t)$, $a(t)$ and d_t so that **PrOnSCP** will keep the server idling for some time and then turn it off, but **OffSCP** will turn off the server at the beginning of the I_1. Suppose **PrOnSCP** keeps the server idling for δ slots given no workload within the look-ahead window and then turn it off. Then according to Algorithm 4.4, we must have $\sum_{\delta+w} d_t p(t) < \beta_s$ and $\sum_{\delta+w+1} d_t p(t) \geq \beta_s$. In this case, $\text{Cost}_{I_1}(\bar{x}) = \beta_s$ and

$$\text{Cost}_{I_1}(x^{\text{on}}) = \sum_{\delta} d_t\, p(t) + \beta_s + \text{c}$$

$$= \sum_{\delta + \omega} d_t\, p(t) - \sum_{\omega} d_t\, p(t) + \beta_s + \text{c}$$

$$\leq \beta_s - d_{\min} P_{\min}\omega + \beta_s + \text{c}$$

$$= \beta_s \left(2 - \frac{d_{\min} P_{\min}}{\beta_s}\omega\right) + \text{c}.$$

So

$$\frac{C_{\text{CP}}(x^{\text{on}})}{C_{\text{CP}}(\bar{x})} \leq \frac{\text{Cost}_{I_1}(x^{\text{on}})}{\text{Cost}_{I_1}(\bar{x})}$$

$$\leq \frac{\beta_s\left(2 - \frac{d_{\min} P_{\min}}{\beta_s}\omega\right) + \text{c}}{\beta_s + \text{c}} \leq 2 - \frac{d_{\min} P_{\min}}{\beta_s}\omega.$$

Combining the above two cases establishes this lemma.

Furthermore, we have some important observations on x^{on} and $\bar{x}$, which will be used in later proofs.

$$\sum_{t=1}^{T} \left[x^{\text{on}}(t) - x^{\text{on}}(t-1)\right]^{+} = \sum_{t=1}^{T} \left[\bar{x}(t) - \bar{x}(t-1)\right]^{+}. \tag{A.5}$$

This is because during an I_1 with $\sum_{t \in I_1} d_t\, p(t) \geq \beta_s$, x^{on} keeps the server idling for some time and then turn it off. $\bar{x}$ turns off the server at the beginning of the I_1. Both x^{on} and $\bar{x}$ turn on the server at the beginning of the following I_2. During an I_1 with $\sum_{t \in I_1} d_t\, p(t) < \beta_s$, both x^{on} and $\bar{x}$ keep the server idling till the following I_2. Thus, x^{on} and $\bar{x}$ incur the same server switching cost. Besides, in both above cases, $x^{\text{on}}(t)$ is no less than $\bar{x}(t)$, we have

$$x^{\text{on}} \geq \bar{x}. \tag{A.6}$$

We also observe that

$$\sum_{t=1}^{T} d_t\, p(t)\left(x^{\text{on}}(t) - \lceil a(t) \rceil\right)$$

$$\leq \sum_{t=1}^{T} d_t\, p(t)\left(\bar{x}(t) - \lceil a(t) \rceil\right) +$$

$$(1 - \alpha_s) \sum_{t=1}^{T} \left[\bar{x}(t) - \bar{x}(t-1)\right]^{+}. \tag{A.7}$$

By rearranging the terms, we obtain

$$\sum_{t=1}^{T} d_t p(t) \left(x^{\mathbf{on}}(t) - \bar{x}(t) \right) \leq (1 - \alpha_s) \sum_{t=1}^{T} [\bar{x}(t) - \bar{x}(t-1)]^{+} . \tag{A.8}$$

Notice that $\sum_{t=1}^{T} d_t p(t) (x(t) - \lceil a(t) \rceil)$ can be seen as the total server idling cost incurred by solution x. Since idling only happens in I_1, Eq. (A.7) follows from the cases discussed above. $\square$

Lemma A.2 $(2 - \alpha_s)$ *is the lower bound of competitive ratio of any deterministic online algorithm for problem* **SCP** *and also* **SCP***, where* $\alpha_s \triangleq \min (1, w d_{\min} P_{\min}/\beta_s) \in [0, 1]$.

Proof First, we show this lemma holds for problem **SCP**. We distinguish two cases:

Case 1: $\omega \geq \beta_s/(d_{\min} P_{\min})$. In this case,$(2 - \alpha_s) = 1$, which is clearly the lower bound of competitive ratio of any online algorithm.

Case 2: $\omega < \beta_s/(d_{\min} P_{\min})$. Similar as the proof of Lemma A.1, we only need to analyze behaviors of online and offline algorithms during an *idle interval* I_1.

Consider the input: $d_t = d_{\min}$ and $p(t) = P_{\min}, \forall t \in [1, T]$. Under this input, during an I_1, we only need to consider a set of deterministic online algorithms with the following behavior: either keep the server idling for the whole I_1 or keep it idling for some slots and then turn if off until the end of the I_1. The reason is that any deterministic online algorithm not belonging to this set will turn off the server at some time and turn on the server before the end of I_1, and thus there must be an online algorithm incurring less cost by turning off the server at the same time but turning on the server at the end of I_1.

We characterize an algorithm **ALG** belonging to this set by a parameter δ, denoting the time it keeps the server idling for given $a \equiv 0$ within the lookahead window. Denote the solutions of algorithms **ALG** and **OffSCP** for problem **SCP** to be $x^{\mathbf{ALG}}$ and $\bar{x}$, respectively.

If δ is infinite, the competitive ratio is apparently infinite due to the fact that the adversary can construct an I_1 whose duration is infinite. Thus we only consider those algorithms with finite δ. The adversary will construct inputs as follows:

If $\delta + w \geq \beta_s/(d_{\min} P_{\min})$, the adversary will construct an I_1 whose duration is longer than $\delta + w$. In this case, **ALG** will keep server idling for δ slots and then turn if off while **OffSCP** turns off the server at the beginning of the I_1 (c.f. Fig. A.1). Then the ratio is

$$
\begin{aligned}
\frac{C_{\text{CP}}(x^{\mathbf{ALG}})}{C_{\text{CP}}(\bar{x})} &= \frac{\sum_{\delta} d_{\min} P_{\min} + \beta_s + d_{\min} P_{\min}}{\beta_s + d_{\min} P_{\min}} \\
&> 1 + \frac{[\beta_s/(d_{\min} P_{\min}) - w] d_{\min} P_{\min}}{\beta_s + d_{\min} P_{\min}} \\
&= 2 - \frac{d_{\min} P_{\min}(\omega + 1)}{\beta_s + d_{\min} P_{\min}} .
\end{aligned}
$$

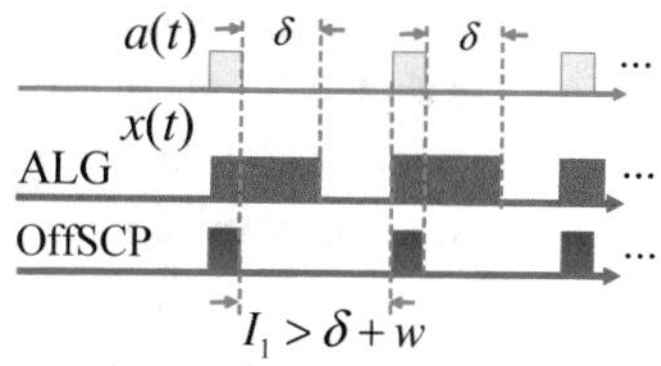

Fig. A.1 A worst case example under the setting that $\delta + \omega \geq \beta_s/(d_{\min} P_{\min})$

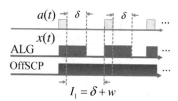

Fig. A.2 Another worst case example under the setting that $\delta + \omega < \beta_s/(d_{\min} P_{\min})$

If $\delta + \omega < \beta_s/(d_{\min} P_{\min})$, the adversary will construct an I_1 whose duration is exactly $\delta + \omega$. In this case, **ALG** will keep server idling for δ slots and then turn if off while **OffSCP** keeps the server idling during the whole I_1 (c.f. Fig. A.2). Then the ratio is

$$
\begin{aligned}
\frac{C_{CP}(x^{\mathbf{ALG}})}{C_{CP}(\bar{x})} &= \frac{\sum_\delta d_{\min} P_{\min} + \beta_s + d_{\min} P_{\min}}{d_{\min} P_{\min}(\delta + \omega) + d_{\min} P_{\min}} \\
&= \frac{d_{\min} P_{\min}(\delta + \omega + 1) + \beta_s - \omega d_{\min} P_{\min}}{d_{\min} P_{\min}(\delta + \omega + 1)} \\
&\geq 1 + \frac{\beta_s - \omega d_{\min} P_{\min}}{\beta_s + d_{\min} P_{\min}} \\
&= 2 - \frac{d_{\min} P_{\min}(\omega + 1)}{\beta_s + d_{\min} P_{\min}}.
\end{aligned}
$$

When $d_{\min} \to 0$ or $\beta_s \to \infty$, we have

$$
2 - \frac{d_{\min} P_{\min}(\omega + 1)}{\beta_s + d_{\min} P_{\min}} \to 2 - \frac{d_{\min} P_{\min}\omega}{\beta_s}.
$$

Combining the above two cases establishes the lower bound for problem **SCP**.

For problem **SCP**, consider the case that $d_t(0) = 0$ and $a(t) \in [0, 1]$, $\forall t$. Thus, the lower bound for **SCP** is also a lower bound for **SCP**. □

Theorem 4.4 follows from Lemmas A.1 and A.2.

Theorem A.2 *RPrOnSCP (Algorithm 4.7) for problem **SCP** has a competitive ratio of $\frac{e}{e-1+\alpha_s}$, where $\alpha_s = \min\{1, \omega d_{\min} P_{\min} \frac{1}{\beta_s}\}$. Further, no randomized online algorithm with a prediction window ω can achieve a smaller competitive ratio.*

Similar as the proof of Theorem 4.4, we only need to focus on the idling interval I_1, $[t_1, t_2]$. Let $d_t \triangleq g_t(1, 0) - g_t(0, 0)$. We let the constant cost $\mathsf{c} = \sum_{t=t_1}^{t_2} p(t) g_t(0, 0)$ which is independent of $x(t)$. For simplicity, we assume $\mathsf{c} = 0$. As we saw in Theorem 4.4, c does not affect the competitive ratio.

RPrOnSCP performs as follows: it accumulates an "idling cost" and when it is less than Λ, it keeps the server idling; otherwise, it will see whether the job will comes, i.e., $a > 0$, before the "idling cost" reaches β_s within the look-ahead window ω. If so, it keeps the server idling; else it turns off the server. Let $\bar{x}$ and $x^{\mathbf{on}}$ be the solutions obtained by **OffSCP** (Algorithm 4.3) and **RPrOnSCP** (Algorithm 4.7), respectively. As shown in the proof of Theorem 4.4, the cost of the offline optimal **OffSCP** is

$$\mathrm{Cost}_{I_1}(\bar{x}) = \begin{cases} D, & \text{if } D < \beta_s, \\ \beta_s, & \text{else,} \end{cases}$$

where $D \triangleq \sum_{t \in I_1} d_t p(t)$.

According to Algorithm 4.7, when $D < \alpha_s \beta_s$, we have

$$\mathbb{E}[\mathrm{Cost}_{I_1}(\bar{x})] = D;$$

when $\alpha_s \beta_s \le D < \beta_s$, we have

$$\mathbb{E}[\mathrm{Cost}_{I_1}(x^{\mathbf{on}})] \le \int_0^{D - \alpha_s \beta_s} (\beta_s + \Lambda) f_\Lambda(\lambda) d\Lambda + \int_{D - \alpha_s \beta_s}^{(1 - \alpha_s)\beta_s} D f_\Lambda(\lambda) d\Lambda;$$

when $D \ge \beta_s$, we have

$$\mathbb{E}[\mathrm{Cost}_{I_1}(x^{\mathbf{on}})] \le \int_0^{(1 - \alpha_s)\beta_s} (\beta_s + \Lambda) f_\Lambda(\lambda) d\Lambda;$$

According to PDF $f_\Lambda(\lambda)$ (Eq. (4.5)), we can calculate $\mathbb{E}[\mathrm{Cost}_{I_1}(x^{\mathbf{on}})]$ and the ratio between $\mathbb{E}[\mathrm{Cost}_{I_1}(x^{\mathbf{on}})]$ and $\mathbb{E}[\mathrm{Cost}_{I_1}(\bar{x})]$:

$$\frac{\mathbb{E}[\mathrm{Cost}_{I_1}(x^{\mathbf{on}})]}{\mathrm{Cost}_{I_1}(\bar{x})} \le \begin{cases} 1, & \text{if } D < \alpha_s \beta_s, \\ \frac{e}{e - 1 + \alpha_s}, & \text{else.} \end{cases}$$

So

$$\frac{\mathbb{E}[C_{\mathrm{CP}}(x^{\mathbf{on}})]}{C_{\mathrm{CP}}(\bar{x})} \le \frac{\mathbb{E}[\mathrm{Cost}_{I_1}(x^{\mathbf{on}})]}{\mathrm{Cost}_{I_1}(\bar{x})} \le \frac{e}{e - 1 + \alpha_s}.$$

Hence, according to Theorem , **RPrOnSCP** achieves the same competitive ratio $\frac{e}{e - 1 + \alpha_s}$ for **SCP**.

Next, we prove no randomized algorithm can achieve a smaller competitive ratio. We set $d_t(x) = d_{min}x$, $p(t) = P_{min}$, $\forall t \in [1, T]$.

Lemma A.3 $\frac{e}{e-1+\alpha_s}$ *is the lower bound of competitive ratio of any randomized online algorithm for problem* **SCP** *and also* **SCP**, *where* $\alpha_s \triangleq \min(1, wd_{min}P_{min}/\beta_s) \in [0, 1]$.

Proof Consider the case that the server becomes empty at τ_1 and it will receive its next job at τ_2. In order to find the best competitive ratio for a randomized online algorithm, it is sufficient to find the minimal ratio of the cost by a randomized online algorithm to that of the offline optimal in $[\tau_1, \tau_2]$. The competitive ratio cannot be lower than the competitive ratio on an instance with a single empty interval, and so we consider that case. We first divide time period (τ_1, τ_2) into slots of equal length. As the length of the slots goes to zero, we can get the best competitive ratio for a continuous time randomized online algorithm.

Assume the critical interval Δ contains exact b slots and there are D slots in $[\tau_1, \tau_2]$. We focus on the case that the look-ahead window has $k \leq b - 2$ slots. (If $k \geq b - 1$, the online algorithm can achieve the offline optimum and the competitive ratio is 1.) Let p_i denote the probability that the algorithm decides to turn off the server at slot $i = 1, 2, \ldots$. Let the competitive ratio be c. Regardless of the value of D, the expected online cost must be at most the competitive ratio times the off-line cost. Thus the minimum competitive ratio satisfies

$$\inf c \tag{A.9}$$

$$\text{s.t. } D\sum_{i=1}^{\infty} p_i \leq cD, \quad \forall D \in [0, k], \tag{A.10}$$

$$\sum_{i=1}^{D-k} (b+i-1)\, p_i + \sum_{i=D-k+1}^{\infty} Dp_i \leq Dc, \quad \forall D \in (k, b] \tag{A.11}$$

$$\sum_{i=1}^{D-k} (b+i-1)\, p_i + \sum_{i=D-k+1}^{\infty} Dp_i \leq bc, \quad \forall D \in (b, \infty] \tag{A.12}$$

$$\sum_{i=1}^{\infty} p_i = 1, \quad 0 \leq p_i \leq 1, \forall i \tag{A.13}$$

$$\text{var } c, p_i, \quad \forall i \in \{1, 2, 3, \ldots\} \tag{A.14}$$

We can show that the optimal value c_d^* of problem (A.9)–(A.14) is equal to the optimal value $\bar{c}^*$ of following problem.

$$\min \bar{c} \qquad (A.15)$$

$$\text{s.t. } 1 \leq \bar{c}, \quad \forall D \in [0, k], \qquad (A.16)$$

$$\sum_{i=1}^{D-k} (b+i-1)\, \bar{p}_i + \sum_{i=D-k+1}^{b-k} D\bar{p}_i \leq D\bar{c}, \quad \forall D \in (k, b) \qquad (A.17)$$

$$\sum_{i=1}^{b-k} (b+i-1)\, \bar{p}_i \leq b\bar{c}, \quad \forall D \in [b, \infty] \qquad (A.18)$$

$$\sum_{i=1}^{b-k} \bar{p}_i = 1 \quad 0 \leq p_i \leq 1, \forall i \qquad (A.19)$$

$$\text{var } \bar{c}, \bar{p}_i, \quad \forall i \in \{1, 2, \ldots, b-k\} \qquad (A.20)$$

Next, we prove that $\bar{p}_1^*$ is positive. If instead $\bar{p}_1^* = 0$, let j be the minimal i such that $\bar{p}_i^* > 0$. Then the constraints (A.17)–(A.18) must hold as strict inequalities for $D \leq k + j - 1$, for the following reason. First consider the constraint for $D = k + j$. Since we have $j \leq b - k - 1$ (otherwise we obtain the deterministic algorithm CSR, which is suboptimal), we have $D = k + j < b$ and the constraint for $D = k + j$, divided by D, is

$$\frac{b+j-1}{k+j} \bar{p}_j + \sum_{i=j+1}^{b-k} \bar{p}_i \leq \bar{c}$$

and when $D \leq k + j - 1$, the constraints, divided by D, are

$$\sum_{i=j}^{b-k} \bar{p}_i \leq \bar{c}.$$

Since $k \leq b - 2$, if the latter were active, then the former would be violated.

We use the slackness of these constraints to show $\bar{p}_1^* > 0$. The coefficient of $\bar{p}_1^*$ is less than that of $\bar{p}_j^*$ in the constraints for $D > k + j - 1$. Therefore, we can decrease $\bar{p}_j^*$ a little bit and increase $\bar{p}_1^*$ a little bit such that all the constraints of (A.16)–(A.19) have slackness, which means we can find a smaller $\bar{c}$ which satisfies all the constraints. This contradicts the optimality of $\bar{p}^* = [\bar{p}_1^*, \bar{p}_2^*, \bar{p}_3^*, \cdots, \bar{p}_{b-k}^*]$. Therefore, we must have $\bar{p}_1^* > 0$.

Next, we show that each of the inequalities in (A.17), (A.18) is tight. Assume instead that the constraint corresponding to some particular $D \in (k, b]$ is loose. Let $D^\#$ be the largest such D. Consider case (i) that $D^\# < b$. Note that $\bar{p}_{D^\#-k+1}^* > 0$, since otherwise $D^\# + 1$ would also be slack. Then decrease $\bar{p}_{D^\#-k+1}^*$ and increase $\bar{p}_{D^\#-k}^*$ slightly. This does not affect constraints for smaller D, but introduces slack into the constraints for all larger D. Next, we could increase $\bar{p}_{D^\#-k}^*$ and decrease $\bar{p}_1^*$, which doesn't affect constraints for larger D, but introduces slack for all constraints with smaller D. Alternatively, in case (ii) that $D^\# = b$, we can decrease $\bar{p}_1^*$ while increasing $\bar{p}_{b-k}^*$ to introduce slack into earlier constraints.

In either case, the transformation induces slack in all constraints, which allows $\bar{c}^*$ to decrease, contradicting the optimality of $\bar{c}^*$. Therefore, all the constraints for $D \in (k, b]$ must be tight.

Since the total $b - k$ constraints for all the $D \in (k, b]$ is tight and $\sum_{i=1}^{b-k} \bar{p}_i = 1$, we can solve the system of linear equations and get the minimal competitive ratio and probability distribution:

$$\bar{c}^* = \left(1 - \left(\frac{b-k-1}{b-k}\right)^{b-k-1} \frac{b-k-1}{b}\right)^{-1}$$

$$\bar{p}^*_{b-k-i} = \frac{\bar{c}^*}{b-k} \left(\frac{b-k-1}{b-k}\right)^i, \quad 0 \le i < b-k-1$$

$$\bar{p}^*_1 = \left(\frac{b-k-1}{b-k}\right)^{b-k-1} \frac{k+1}{b} \bar{c}^*, \quad k < b$$

Letting b go to infinity and keeping $k/b = \alpha$, we have the minimal competitive ratio c^* for continuous time:

$$c^* = \frac{e}{e - 1 + \alpha}$$

This means the optimal competitive ratio for continuous time randomized online algorithm is $c^* = \frac{e}{e-1+\alpha}$, as required. Therefore, the best competitive ratio of randomized algorithm for single ski-rental problem is $\frac{e}{e-1+\alpha}$. We have the following lemma to prove that **RPrOnSCP** has the best competitive ratio for randomized algorithms against an oblivious adversary.

In the proof we will use the notation used in the proof of lemma 13. Assume that the cost of online algorithm in the ith ski rental is C and the corresponding offline optimal is C^*. If the strategy of the oblivious adversary is to arbitrarily choose a number which is greater than $\alpha\Delta$ as the empty period of each ski rental problem, then the online algorithm has no information of the length of current empty period $T_{i,E}$ even if the online algorithm knows $T_{i-1,E}, T_{i-2,E},...T_{1,E}$. We are going to prove lemma 16 by induction.

It is clear that in the first ski rental we can not do better than in single ski rental problem. Therefore, we have $\mathbb{E}(C_1) \le \frac{e}{e-1+\alpha} C_1^*$.

Assume that $\mathbb{E}\left(\sum_{i=1}^{k-1} C\right) \le \frac{e}{e-1+\alpha} \sum_{i=1}^{k-1} C^*$, we are going to prove $\mathbb{E}\left(\sum_{i=1}^{k} C\right) \le \frac{e}{e-1+\alpha} \sum_{i=1}^{k} C^*$.

Suppose online algorithm chooses $f_{Z_k}(z_k|z_{k-1}, z_{k-2}...z_1)$ as the conditional probability distribution of Z_k given the historical information. Then there always exists a $\bar{T}_{k,E}$ such that

$$\mathbb{E}(C_k) = \mathbb{E}(\mathbb{E}(C_k|Z_{k-1}, Z_{k-2}...Z_1))$$

$$\ge \frac{e}{e-1+\alpha} C_k^*$$

To see this, suppose there is no $\bar{T}_{k,E}$ satisfying above inequality, then for any $T_{k,E}$, we must have

$$\mathbb{E}\left(\mathbb{E}\left(C_k | Z_{k-1}, Z_{k-2}...Z_1\right)\right) < \frac{e}{e-1+\alpha} C_k^*$$

Then in the single ski rental problem, we can also let the distribution of random variable Z follow the unconditional distribution $f_{Z_k}(z_k)$ of Z_k. In this way, we can get a better competitive ratio than $\frac{e}{e-1+\alpha}$ for a single ski rental problem. This is a contradiction. Therefore, such $\bar{T}_{k,E}$ must exist. This means the best ratio we can do in kth ski rental is $\frac{e}{e-1+\alpha}$. Since

$$\mathbb{E}\left(\sum_{i=1}^{k} C\right) = \mathbb{E}\left(\sum_{i=1}^{k-1} C\right) + \mathbb{E}\left(C_k\right) \text{ and } \sum_{i=1}^{k} C^* = \sum_{i=1}^{k-1} C^* + C_k^*, \text{ thus we have}$$

$$\mathbb{E}\left(\sum_{i=1}^{k} C\right) \leq \frac{e}{e-1+\alpha} \sum_{i=1}^{k} C^*$$

as required. This means online algorithms can not do better than $\frac{e}{e-1+\alpha}$ even against oblivious adversaries. Therefore, **RPrOnSCP** has the best competitive ratio $\frac{e}{e-1+\alpha}$ for randomized algorithms against oblivious adversary. Since **SCP** is a repeated ski rental problem as a special case, $\frac{e}{e-1+\alpha}$ is the optimal competitive ratio for any randomized algorithm. $\square$